# Sleep Research

# Sleep Research

Proceedings of the Northern European Symposium
on Sleep Research
Basle, September 26 and 27, 1978

Edited by

R. G. Priest

*Professor of Psychiatry, St Mary's Hospital
Medical School, University of London*

A. Pletscher

*Director, Research Division,
F. Hoffmann-La Roche & Co. Ltd.,
Basle, Switzerland*

and

J. Ward

*Department of Medical Information,
International Pharma Division,
F. Hoffmann-La Roche & Co. Ltd.,
Basle, Switzerland*

**MTP**PRESS LIMITED
*International Medical Publishers*

All names of products in this book
are legally protected as Trade Marks.

Published by
MTP Press Limited
Falcon House
Lancaster, England

British Library Cataloguing in Publication Data

Northern European Symposium on Sleep Research,
   *Basle, 1978*
   Sleep research.
   1. Sleep–Congresses
   I. Title   II. Priest, Robert G.   III. Pletscher, A.
   IV. Ward, John
   612'.821       QP425

ISBN-13: 978-94-011-6228-9     e-ISBN-13: 978-94-011-6226-5
DOI: 10.1007/978-94-011-6226-5

# Contents

List of Contributors                                                    ix

Opening Remarks
*A. Pletscher*                                                          xiii


## Section I
## Physiology and Pathophysiology of Sleep
*Moderator: Professor R. G. Priest*

**1** The benzodiazepine receptor in human brain
*H. Möhler and T. Okada*                                                 3

**2** Cerebral circulation and metabolism in sleep
*D. H. Ingvar*                                                          13

**3** α- and β-adrenergic mechanisms in the control of sleep
stages
*P. T. S. Putkonen*                                                     19

**4** Brain catecholaminergic activity in relation to sleep
*J.-M. Gaillard*                                                        35

**5** Melatonin and sleep in man: a preliminary report
*T. Hansen, A. J. Birkeland and O. Lingjærde*                           43

Discussion I                                                           45


## Section II
## Treatment of Sleep Disorders
## 1. Parameters of Efficacy
*Moderator: Professor J. Bastiaans*

**6** Clinical pharmacokinetic and biopharmaceutical aspects
of hypnotic drug therapy
*D. D. Breimer*     63

**7** Clinical and psychometric effects of flunitrazepam
observed during the day in relation to
pharmacokinetic data
*R. Amrein, J. P. Cano, D. Hartmann, W. H. Ziegler
and R. Dubuis*     83

**8** Correlation of plasma concentrations of benzodiazepines
with clinical effects
*M. Lader*     99

**9** Effects of benzodiazepines on sleep and on performance:
studies in healthy man
*A. N. Nicholson, R. G. Borland and B. M. Stone*     109

**10** Amnesic action and residual effects of benzodiazepines
used for intravenous sedation
*K. Korttila*     123

**11** Sleep and mood: measuring the sleep quality
*P. Visser, W. F. Hofman, A. Kumar, R. Cluydts,
I. P. F. de Diana, P. Marchant, H. J. Bakker, R. van Diest
and P. A. M. Poelstra*     135

    Discussion II     146

**Section III**
**Treatment of Sleep Disorders**
**2. Clinical Experience**
*Moderator: Professor I. Oswald*

**12** Comparative studies with hypnotics
*E. Wickström*     155

**13** Pharmacological treatment of sleep disorders
*C. G. Gottfries*     171

**14** Role of the sleep laboratory in the evaluation of
hypnotic drugs
*C. R. Soldatos and A. Kales*                                    181

**15** A sleep laboratory in a department of psychiatry
*W. Lehmann*                                                    197

Discussion III                                                   203

Index                                                            221

# List of Contributors

**R. AMREIN**
Department of Clinical Investigation
and Development,
Pharma Division,
F. Hoffmann-La Roche & Co. Ltd.,
Basle, Switzerland

**H. J. BAKKER**
Psychophysiology Laboratory,
University of Amsterdam,
Netherlands

**J. BASTIAANS**
The Jelgersmakliniek,
Oegstgeest, Netherlands

**A. J. BIRKELAND**
Institutt for Samfunnsmedisin,
Universitetet i Tromsø,
Asgård Sykehus, Norway

**R. G. BORLAND**
Royal Air Force Institute of Aviation
Medicine,
Farnborough,
Hampshire, United Kingdom

**D. D. BREIMER**
Department of Pharmacology,
Subfaculty of Pharmacy,
University of Leiden,
Leiden, Netherlands

**J. P. CANO**
Faculty of Pharmacy,
Marseille, France

**R. CLUYDTS**
Laboratory of Physiology and
Physiopathology,
Free University of Brussels,
Belgium

**I. P. F. de DIANA**
Laboratory of Physiology and
Physiopathology,
Free University of Brussels,
Belgium

**R. van DIEST**
Psychophysiology Laboratory,
University of Amsterdam,
Netherlands

**R. DUBUIS**
Department of Clinical Investigation
and Development,
Pharma Division,
F. Hoffmann-La Roche & Co. Ltd.,
Basle, Switzerland

**J.-M. GAILLARD**
Clinique Psychiatrique de l'Université
de Genève,
Bel-Air,
Chêne-Bourg, Switzerland

**C. G. GOTTFRIES**
University Department of Psychiatry,
Sahlgrenska sjukhuset,
Göteborg, Sweden

**T. HANSEN**
Institutt for Samfunnsmedisin,
Universitetet i Tromsø,
Asgård Sykehus, Norway

**D. HARTMANN**
Department of Clinical Investigation
and Development,
Pharma Division,
F. Hoffmann-La Roche & Co. Ltd.,
Basle, Switzerland

ix

**W. F. HOFMAN**
Psychophysiology Laboratory,
University of Amsterdam,
Netherlands

**D. H. INGVAR**
Department of Clinical Neuro-
   physiology,
University Hospital,
Lund, Sweden

**A. KALES**
Department of Psychiatry,
Sleep Research and Treatment Center,
Pennsylvania State University College
   of Medicine,
Hershey, Pennsylvania, USA

**K. KORTTILA**
Department of Anaesthesia,
Helsinki University Central Hospital,
Helsinki, Finland

**A. KUMAR**
Psychophysiology Laboratory,
University of Amsterdam,
Netherlands

**M. LADER**
Institute of Psychiatry,
University of London,
United Kingdom

**W. LEHMANN**
Department of Psychiatry,
University Hospital,
Uppsala, Sweden

**O. LINGJÆRDE**
Institutt for Samfunnsmedisin,
Universitetet i Tromsø,
Asgård Sykehus, Norway

**P. MARCHANT**
Laboratory of Physiology and
   Physiopathology,
Free University of Brussels,
Belgium

**H. MOHLER**
Department of Clinical Investigation
   and Development,
Pharma Division,
F. Hoffmann-La Roche & Co. Ltd.,
Basle, Switzerland

**A. N. NICHOLSON**
Royal Air Force Institute of Aviation
   Medicine,
Farnborough,
Hampshire, United Kingdom

**T. OKADA**
Biochemistry Department,
Nippon-Roche Research Centre,
200 Kajiwara,
Kamakura-City, Japan

**I. OSWALD**
University Department of Psychiatry,
Royal Edinburgh Hospital,
Scotland

**A. PLETSCHER**
Research Division,
F. Hoffmann-La Roche & Co. Ltd.,
Basle, Switzerland

**P. A. M. POELSTRA**
Psychophysiology Laboratory,
University of Amsterdam,
Netherlands

**R. G. PRIEST**
St Mary's Hospital Medical School,
University of London,
United Kingdom

**P. T. S. PUTKONEN**
Department of Physiology,
University of Helsinki,
Helsinki, Finland

**C. R. SOLDATOS**
Department of Psychiatry,
Sleep Research and Treatment Center,
Pennsylvania State University College
   of Medicine,
Hershey, Pennsylvania, USA

**B. M. STONE**
Royal Air Force Institute of Aviation
   Medicine,
Farnborough,
Hampshire, United Kingdom

**P. VISSER**
Psychophysiology Laboratory,
University of Amsterdam,
Netherlands

**J. WARD**
Department of Medical Information,
International Pharma Division,
F. Hoffmann-La Roche & Co. Ltd.,
Basle, Switzerland

**E. WICKSTRØM**
Ullevål sykehus,
Oslo, Norway

**W. H. ZIEGLER**
Department of Clinical Investigation
 and Development,
Pharma Division,
F. Hoffmann-La Roche & Co. Ltd.,
Basle, Switzerland

# Opening Remarks

## A. Pletscher

Ladies and gentlemen, welcome to Roche Basle. It is a great pleasure and honour for me to open this symposium on sleep research, and I thank all of you who have kindly agreed to participate in this event.

Sleep is certainly one of the most fascinating phenomena in biology. Sleep disturbances due to various causes are amongst the most common disorders of the central nervous system of man. Therefore, this phenomenon has attracted the curiosity of scientists for centuries and research on the physiology and pathophysiology of sleep can look back on a very long tradition. The progress in sleep research of the last 50 years has been especially remarkable.

I was fortunate enough to be a student of W. R. Hess in Zurich, who was awarded the Nobel prize for his basic studies on new concepts of the role of the central nervous system in regulating autonomic fuctions, including sleep. The success of Rudolf Hess was not based only on his ingenious electrophysiological techniques, which were anyway rather primitive for today's standards, but also on his meticulous observations of the animals' behaviour.

Since Hess's classical work, electrophysiological techniques have developed impressively and for clinical studies the application of the EEG has been most fruitful. Since the late 'forties of this century another aspect of sleep research has developed, namely the biochemistry, neurophysiology and anatomy of neurohumoral transmitters. Through the classical work of Dale, Loewy, Euler and others, acetylcholine and norepinephrine were found to be mediators of the peripheral autonomic functions. These substances, together with 5-hydroxytryptamine, serotonin, were then also discovered to occur in the brain with a specific distribution, to be synthesized, stored, released and metabolized in this complex organ.

With ingenious biochemical, electrophysiological and morphological techniques it was found that these substances functioned as neurohumoral transmitters and that their turnover could be influenced by various drugs. Other types of transmitters were then discovered and today the number of established or putative neurotransmitters and neuromodulators in the central nervous system exceeds 20; almost every day now a new transmitter – be it a peptide or a low molecular weight substance – is being discovered. Among these transmitters are GABA, glutamic acid, polypeptides like the endorphins and enkephalins to name only a few. Good evidence exists that several of these compounds are involved in the physiology and pathophysiology of sleep. The first of these transmitters to be discovered to be involved in sleep was probably 5-hydroxytryptamine, which is concentrated in the raphe nuclei.

Roche has had a long interest in sleep disorders and sleep research. This company was probably one of the first to develop new nonbarbiturate hypnotics like Persedon and Noludar. Roche also did pioneer work in the field of biogenic amines. In this context Dr Markus Guggenheim must be mentioned, a former research director of Roche, who started such work in the early 'twenties of this century, and whose book on biogenic amines was until recently a classic.

The latest developments of Roche in the treatment of sleep disorders were the benzodiazepines, one of the topics of today's symposium. The mechanism of action of these compounds has remained obscure for a long time and is still not yet fully elucidated. However, exciting new developments have taken place in the last few years. The mechanism of action of the benzodiazepines seems to involve $\gamma$-aminobutyric acid and the GABA-ergic system, and the benzodiazepines have become a most valuable tool for elucidating the physiology and pathophysiology of this interesting brain system.

It also seems possible that this GABA-ergic system is somehow involved in the mechanism of sleep. Thus, on the one hand, much progress has been made in recent years, but on the other hand the mechanism of sleep is still far from being fully elucidated and remains a further challenge for neurophysiologists, pharmacologists and clinicians.

It seems therefore to be an appropriate time to hold a symposium on sleep research. It is the aim of this symposium to exchange and review the presently available experience on sleep research with a view to both theoretical and practical aspects. In particular the mechanisms of sleep and its disturbances will be discussed and emphasis will be put on the pharmacological treatment of sleep disturbances and on comparisons of the therapeutic usefulness of such treatments. We are looking forward to an interesting and stimulating meeting, and perhaps also some controversial discussions.

# Section I

# Physiology and Pathophysiology of Sleep

Moderator: Professor R. G. Priest

# 1

# The Benzodiazepine Receptor in Human Brain

## H. Möhler and T. Okada

## 1.1 INTRODUCTION

Benzodiazepines have found wide therapeutic application in the treatment of anxiety, sleep disorders, muscle spasms and convulsions. In the present report we describe recent progress in the clarification of the mechanism of action of benzodiazepines: benzodiazepines are bound to a specific target structure in the brain, termed benzodiazepine receptor, in order to elicit their pharmacological and therapeutic central effects. The receptor is characterized by a binding site highly specific for benzodiazepines. In addition to the identification, characteristics and distribution of the benzodiazepine receptor in normal human brain, its alteration in Huntington's disease are described. Furthermore, present views on the biochemical mechanism of action of benzodiazepine are discussed.

In analogy to the studies in rat brain[3,14,15,25], a benzodiazepine binding site in human brain was identified by equilibrium binding studies *in vitro*, using the particulate fraction of previously frozen postmortem human brain (1 mg protein) and $^3$H-diazepam (usually 1.5 nmol) as labelled ligand. The amount of labelled ligand which could be displaced from the binding site by a high concentration of an unlabelled potent benzodiazepine (e.g. 1 $\mu$mol diazepam) was termed specifically bound. In order to establish that specific binding of the labelled ligand represents an interaction with a pharmacologically meaningful binding site, several criteria have to be fulfilled:

1. There should be only a limited number of specific binding sites in the tissue material.
2. The interaction of various benzodiazepines with the binding site should reflect their pharmacological potency.
3. An uneven distribution of the specific binding site in the brain would be expected.
4. Assuming an involvement in synaptic events, the binding site should be enriched in the synaptic-membrane fraction.

Postmortem brain tissue from control patients who died of various heart diseases with no disorder of the central nervous system (CNS) was obtained from the Institute of Pathology, Basle. Huntington's chorea brain tissue and the respective controls were kindly provided by Dr E. Bird and Dr E. G. Spokes, Addenbrooke's Hospital, Cambridge, England.

## 1.2 HIGH-AFFINITY BINDING SITE FOR BENZODIAZEPINES

Pharmacologically potent benzodiazepines show a high apparent affinity to the binding site, as indicated by the low concentration of these compounds inhibiting $^3$H-diazepam-specific binding by 50% (Figure 1.1). Correspondingly, the pharmacologically active enantiomer of a benzodiazepine has a 1000-fold higher displacing potency than the pharmacologically inactive enantiomer (Figure 1.2).

The apparent affinity of various benzodiazepines to the benzodiazepine receptor in human brain correlates very well with their pharmacological and therapeutic potency, as shown for their muscle-relaxant and anticonvulsant effect in animals and their tranquillizing and hypnotic action in man (Table 1.1). Thus, the benzodiazepine-specific binding site appears to be the primary target structure in human brain to which benzodiazepines are bound in order to elicit their therapeutic effects[4,17].

The saturability of the benzodiazepine-specific binding site with increasing concentrations of labelled ligand indicates that the number of binding sites is limited (Figure 1.3).

4

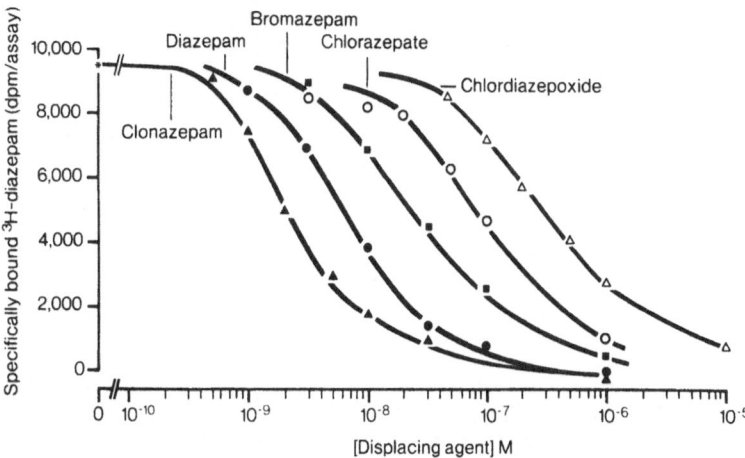

**Figure 1.1** Potency of different benzodiazepines in displacing specifically bound ³H-diazepam. Homogenates of human frontal cerebral cortex were incubated with 1.5 nmol ³H-diazepam and increasing concentrations of various benzodiazepines. The points are the means of triplicate determinations with SEM < 3%. The $K_i$-values from three different experiments are given in Table 1.1

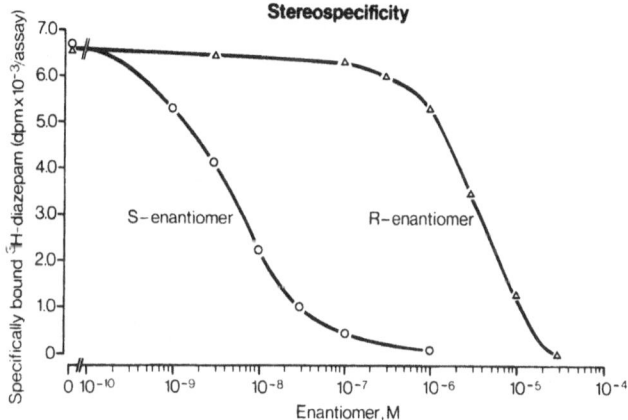

**Figure 1.2** Stereospecificity of ³H-diazepam-specific binding. Homogenates from human frontal cerebral cortex (1 mg protein) were incubated at 4 °C for 15 min in 2 ml Krebs–Ringer–Tris-buffer pH 7.4 containing 1.5 nmol ³H-diazepam and increasing concentrations of the S- or R-enantiomer of the benzodiazepine 5-(ο-fluorophenyl)-1,3-dihydro-1,3-dimethyl-7-nitro-2H-1,4-benzodiazepine-2-one. The points are the means of triplicate determinations with SEM < 3%. The $K_i$-values obtained from three different brains are $K_i = 4.3 \pm 0.2$ nmol for the (+) enantiomer (S-enantiomer) and $K_i = 2800 \pm 30$ nmol for the (−) enantiomer (R-enantiomer). The S-enantiomer is pharmacologically very potent in contrast to the inactive R-enantiomer. For details, see reference 17

5

**Table 1.1 Benzodiazepines: comparison of their affinity to the benzodiazepine binding site with their therapeutic and pharmacological potency**

| Benzodiazepine | Inhibition of $^3$H-diazepam binding in cerebral cortex of | | Antagonism of pentetrazol-induced convulsant in mice $ED_{50}$‡ ($\mu mol/kg$) | Muscle-relaxant action in cat $EDmin$‡ ($\mu mol/kg$) | Average therapeutic dose ($\mu mol/day$)§ | Inhibition of electric-shock-induced fighting of mice $ED_{50}$‡ ($\mu mol/kg$) |
|---|---|---|---|---|---|---|
| | Man* $K_i$ (nmol) | Rat† $K_i$ (nmol) | | | | |
| Clonazepam | 0.87 ± 0.07 | 1.5 | 0.95 | 0.16 | 11.7 | 6.3 |
| Flunitrazepam | 2.2 ± 0.5 | 2.8 | 0.32 | 0.06 | 7.9 | 2.5 |
| Lorazepam | 2.3 ± 0.3 | 2.7 | 0.62 | 0.78 | 7.7 | 6.2 |
| Ro 11-7800 | 2.3 ± 0.1 | 1.9 | — | — | — | — |
| Triazolam | 2.4 ± 0.2 | 2.8 | — | — | 3.2 | — |
| Ro 11-6896 (3S) | 4.3 ± 0.2 | 4.8 | — | — | — | — |
| Diazepam | 7.4 ± 0.8 | 6.3 | 7.0 | 0.70 | 26 | 34.9 |
| Nitrazepam | 9.2 ± 0.8 | 6.4 | 2.5 | 0.36 | 27 | 17.8 |
| Flurazepam | 11 ± 1 | 11 | 4.7 | 4.7 | 53 | 47.2 |
| Oxazepam | 19 ± 2 | 14 | 2.4 | 3.5 | 157 | 139.4 |
| Bromazepam | 21 ± 1 | 12 | 2.2 | 0.63 | 57 | 6.3 |
| Chlorazepate | 44 ± 6 | 41 | 5.7 | 1.14 | 64 | 56.8 |
| Chlordiazepoxide | 360 ± 50 | 220 | 23.8 | 6.0 | 134 | 119.0 |
| Medazepam | 880 ± 60 | 600 | 22.8 | 13.0 | 85 | 65.1 |
| Ro 11-6893 (3R) | 2800 ± 300 | 1040 | — | — | — | — |
| Ro 5-2181 | 5300 ± 200 | 3700 | 33.1 | 36.7 | — | 147.0 |
| Correlation with the inhibition of $^3$H-diazepam binding in man | | $r = 0.99$ $p < 0.0001$ | $r = 0.90$ $p < 0.0001$ | $r = 0.88$ $p < 0.0001$ | $r = 0.79$ $p < 0.005$ | $r = 0.75$ $p < 0.005$ |

$r$ = correlation coefficient; significance given as $p$.
(−) indicates no data available.
*The $K_i$-values were determined by incubating homogenates of human frontal cerebral cortex (1 mg protein) in triplicate with 1.5 nmol $^3$H-diazepam and six concentrations of each drug in 2 ml Krebs–Ringer–Tris-buffer pH 7.4, as described in reference 3. The 50% inhibitory concentration, $IC_{50}$, was determined by log-logit analysis. Since the inhibition of $^3$H-diazepam binding was competitive[17], $IC_{50}$ was converted to $k_i$, according to the equation $K_i = IC_{50}/(1 + C \cdot K_D$, where $C$ is the concentration of the radioactive ligand and $K_D$ its dissociation constant. Values are mean ± SEM from 3 different brains.
† Data from Möhler, H. and Okada, T.[14]
‡ Data from Randall, L. O. et al.[2]
§ The doses given are the mean values of the dose ranges recommended for therapeutic use of the drugs as minor tranquillizers and hypnotics in adults,

except clonazepam, which is exclusively used as antiepileptic[12]. For the dose range of triazolam, see Gottschalk, L. A. and Elliott, A. W.[9].

The benzodiazepines are listed in order of their affinity to the benzodiazepine binding site in human frontal cortex. The chemical structures of the benzodiazepines are given in Randall, L. O. et al.[23], except Ro 11-7800 = 9-(aminomethyl)-2-chloro-4-(o-chlorophenyl)-6H-thieno < 3,2-f > -s-triazolo < 4,3-a > < 1,4 > diazepam, Ro 11-6896 and Ro 11-6893, which are the (3S) and (3R) enantiomers of 5-(o-fluorophenyl)-1,3-dihydro-1,3-dimethyl-7-nitro-2H-1,4-benzodiazepine-2-one respectively, and triazolam = 8-chloro-6-(o-chlorphenyl)-1-methyl-4H-s-triazolo-(4,3-a) (1,4)-benzodiazepine.

Compounds with no inhibitory effect, at least up to $10^{-6}$ mol on $^3$H-diazepam binding, include phenobarbital, meprobamate, glycine, strychnine, GABA, (+)-bicuculline-methiodide, morphine, naloxone, serotonin, LSD, dopamine, pimozide, norepinephrine, phentolamine, propranolol.

The high-affinity binding site for benzodiazepines can be demonstrated not only *in vitro* but also *in vivo*. In rat brain it has recently become possible to determine the occupancy of the benzodiazepine receptors by labelled benzodiazepines *in vivo*[5, 28]. This method will make it possible to determine the amount of free drug at the receptor

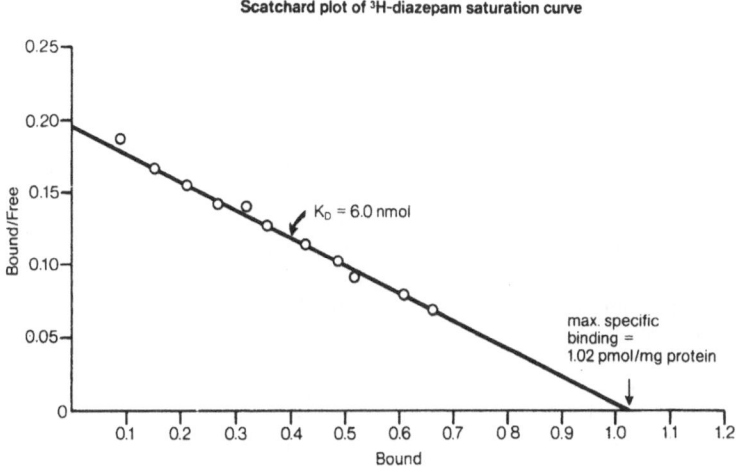

**Figure 1.3** Scatchard plot of ³H-diazepam-specific binding. The amount of specifically bound ligand at various concentrations of ³H-diazepam added to the incubation (0.5 – 10 nmol) was plotted according to Scatchard. Bound = pmol specifically bound ³H-diazepam per mg protein; free = concentration of unbound ³H-diazepam in the incubation medium (nmol). The apparent dissociation constant of diazepam obtained using frontal cerebral cortex from four different brains was $K_D = 7.0 \pm 0.8$ nmol, the maximal specific binding $1.2 \pm 0.2$ pmol/mg protein

site in the brain *in vivo*. Thus, it will be possible to investigate to what extent the plasma concentration of a benzodiazepine reflects the concentration of free drug at the receptor site in the brain.

## 1.3 REGIONAL DISTRIBUTION OF THE BENZODIAZEPINE RECEPTOR

There is a 25-fold variation in the density of the benzodiazepine receptor in human brain, with the highest number of receptors in cortical areas. The affinity of benzodiazepines to the receptor, however, is remarkably similar throughout the brain (Table 1.2).

Thus, the impact of benzodiazepine treatment on neuronal events may be most pronounced in cortical brain areas[4, 17].

**Table 1.2  Maximal specific binding and apparent dissociation constant of diazepam in various regions of human brain**

| Brain region | Maximal specific binding (fmol/mg protein) | $K_D$ (nmol) |
|---|---|---|
| Cerebral cortex | | |
|   Frontal | $1200 \pm 160$ | $7.0 \pm 0.8$ |
|   Precentral | $1200 \pm 200$ | $7.5 \pm 1.9$ |
|   Postcentral | $1110 \pm 60$ | $7.0 \pm 1.0$ |
| Cerebellum | | |
|   Cortex | $730 \pm 70$ | $5.8 \pm 0.8$ |
|   Vermis | $720 \pm 90$ | $8.0 \pm 2.9$ |
| Amygdala | $720 \pm 100$ | $6.7 \pm 1.5$ |
| Hippocampus | $610 \pm 50$ | $6.5 \pm 1.4$ |
| Hypothalamus | $520 \pm 40$ | $8.2 \pm 0.6$ |
| Nucleus accumbens | $430 \pm 40$ | $6.5 \pm 0.8$ |
| Thalamus | $410 \pm 50$ | $8.7 \pm 1.5$ |
| Nucleus caudatus | $380 \pm 60$ | $8.5 \pm 1.3$ |
| Putamen | $360 \pm 60$ | $6.8 \pm 0.9$ |
| Globus pallidus | $300 \pm 10$ | $6.2 \pm 0.8$ |
| Substantia nigra | $290 \pm 50$ | $10.6 \pm 4.1$ |
| Tegmentum | $180 \pm 40$ | $10.1 \pm 3.3$ |
| Dentate nucleus | $160 \pm 30$ | $8.7 \pm 4.3$ |
| Olive | $160 \pm 30$ | $14.1 \pm 2.9$ |
| Pons | $160 \pm 10$ | $14.2 \pm 3.0$ |
| Medulla oblongata | $150 \pm 50$ | $21.9 \pm 9.1$ |
| Corpus callosum | $50 \pm 10$ | $6.4 \pm 1.9$ |

Homogenates of different regions from human brain (1 mg protein) were incubated in triplicate with six concentrations (0.5 to 10 nmol) of $^3$H-diazepam in the presence and absence of unlabelled diazepam (1 µmol), as described in reference 5. Specific binding of $^3$H-diazepam was plotted according to Scatchard in order to obtain the apparent dissociation constant $K_D$ and the value of maximal specific binding. The data are the means $\pm$ SEM from four different brains.

## 1.4  SUBCELLULAR LOCALIZATION OF THE BENZODIAZEPINE RECEPTOR

In subcellular fractionation studies of non-frozen postmortem frontal cortex using differential centrifugation, the benzodiazepine receptor was found to be most enriched in the synaptic-membrane fraction, indicating a possible involvement of the benzodiazepine receptor in synaptic events. This conclusion is in line with a lack of benzodiazepine receptors in white matter, e.g. corpus callosum[21] (see Table 1.2).

## 1.5 CELLULAR LOCALIZATION OF THE BENZODIAZEPINE RECEPTOR

The benzodiazepine receptor appears, at least in some areas of human brain, to be localized on GABAergic neurons. This could be shown in postmortem brain tissue from patients afflicted with Huntington's disease. The degeneration of GABAergic neurons, known to occur in caudate nucleus and putamen[1], was paralleled by a loss of benzodiazepine receptors in these brain areas[16, 20]. This finding could explain why the ameliorative effect of benzodiazepine treatment is not sustained in the later stages of this disease.

It is noteworthy that the loss of benzodiazepine receptors, observed in Huntington's disease, is not paralleled by a loss of GABA receptors, suggesting that the two receptors may not be localized on the same cellular element.

## 1.6 POSSIBLE INVOLVEMENT OF THE BENZODIAZEPINE RECEPTOR IN SYNAPTIC EVENTS

Several lines of evidence suggest that the benzodiazepine receptor may be involved in synaptic events:

1. The benzodiazepine receptor appears to be localized on neurons, since a loss of benzodiazepine receptors is found in Huntington's disease only in those brain areas in which a pronounced loss of neurons in known to occur.
2. Subcellular fractionation studies show that the benzodiazepine receptor is mainly localized in the synaptic membrane fraction.
3. No benzodiazepine receptors could be detected in primary cultures of astroglial cells[2].
4. From electrophysiological experiments it is well known that benzodiazepines enhance GABAergic synaptic transmission[11], suggesting that benzodiazepine receptors may be localized pre-synaptically on GABA neurons or postsynaptically in GABA-ergic synapses.

9

Recently, a postsynaptic localization of the benzodiazepine receptor in GABAergic synapses was postulated in rat brain based on *in vitro* biochemical evidence that benzodiazepines enhance the affinity of the GABA receptor for the neurotransmitter GABA[10, 27]. Similarly, GABA-like compounds enhance the affinity of benzodiazepines to the benzodiazepine receptor[13, 26]. In this view, a close structural association of the GABA and the benzodiazepine receptor would be expected.

However, benzodiazepine receptors and GABA receptors, at least in putamen of human brain, do not appear to be localized on the same cellular element (see above). It appears, therefore, that, at least in some brain areas, benzodiazepine receptors have a function independent of GABA receptors. Thus, two functionally different benzodiazepine receptors may exist in the brain.

It should be kept in mind that benzodiazepines, apart from enhancing GABAergic neurotransmission[11], antagonize excitatory amino acids[19] and acetylcholine[6] and also reduce the turnover of 5-hydroxytryptamine[7, 8]. Benzodiazepine receptors may directly be involved also in these neurotransmitter systems, mediating the effects of benzodiazepines.

## 1.7   POSSIBLE PHYSIOLOGICAL FUNCTION OF THE BENZODIAZEPINE RECEPTOR

The synaptic localization is in line with a possible physiological function of the benzodiazepine receptor in neuronal transmission. In analogy to the discovery of enkephalins and $\beta$-endorphin subsequent to the finding of morphine receptors in the CNS, the brain may contain a compound acting physiologically as a ligand to the benzodiazepine receptor. In this view, the brain would contain its own antianxiety agent.

Nicotinamide, recently isolated from calf brain extracts on the basis of its affinity to the benzodiazepine-receptor binding site, was found to exert benzodiazepine-like effects in all assay systems tested so far, when administered to animals or men[18]. Most relevant to this symposium is the prolongation of total sleep time and the increase in rapid eye movement sleep after nicotinamide application to humans[22, 24], a pattern also seen after low doses of benzodiazepines.

Thus, nicotinamide has to be considered as candidate for the endogenous brain constituent exerting benzodiazepine-like actions in the brain physiologically[18].

## References

1. Bird, E. D. and Iversen, L. L. (1974). Huntington's chorea: postmortem measurement of glutamic acid decarboxylase, choline acetyltransferase and dopamine in basal ganglia. *Brain*, **97**, 457–472
2. Braestrup, C., Nissen, C., Squires, R. F., Schousboe, A. (1970). Lack of brain specific benzodiazepine receptors on mouse primary astroglial cultures. *Neurosci. Lett.*, **9**, 45–49
3. Braestrup, C. and Squires, R. F. (1977). Specific benzodiazepine receptors in rat brain characterized by high-affinity ³H-diazepam binding. *Proc. Natl. Acad. Sci. USA*, **74**, 3805–3807
4. Braestrup, C., Albrechtsen, R. and Squires, R. F. (1977). High densities of benzodiazepine receptors in human cortical areas. *Nature*, **269**, 702–704
5. Chang, R. S. L. and Snyder, S. H. (1978). Benzodiazepine receptors: labelling in intact animals with ³H-flunitrazepam. *Eur. J. Pharmacol.*, **48**, 213–218
6. Davies, J. and Polc, P. (1978). Effect of a water-soluble benzodiazepine on the responses of spinal neurones to acetylcholine and excitatory amino acid analogues. *Neuropharmacology*, **17**, 217–220
7. Feldman, R. (1978). Discriminative properties of chlordiazepoxide: a new method of analysis. *Psychopharmacology*, **58** (2), p. 6, Abstr. No. 13
8. File, S. A. and Vellucci, S. V. (1978). Studies on the role of ACTH and 5-HT in anxiety, using an animal model. *J. Pharm. Pharmacol.*, **30**, 105–110
9. Gottschalk, L. A. and Elliott, A. W. (1976). Effects of triazolam and flurazepam on emotions and intellectual function. *Res. Commun. Psychol., Psychiatr. Behav.*, **1**, 575–595
10. Guidotti, A., Toffano, G. and Costa, E. (1978). An endogenous protein modulates the affinity of GABA and benzodiazepine receptors in rat brain. *Nature*, **275**, 553–555
11. Haefely, W., Polc, P., Schaffner, R., Keller, H. H., Pieri, L. and Möhler, H. (1978). Facilitation of GABAergic transmission. In H. Kofod, P. Krogsgaard-Larsen and F. Scheel-Krüger (eds.), *GABA Neurotransmitters*. (In press)
12. Instructions for medical use provided by the manufacturers on the product information sheet of the commercially available drugs
13. Karobath, M. and Sperk, G. (1979). Stimulation of benzodiazepine receptor binding by γ-aminobutyric acid. *Proc. Natl. Acad. Sci. USA*. (In press)
14. Möhler, H. and Okada, T. (1977). Benzodiazepine receptor: demonstration in the central nervous system. *Science*, **198**, 849–851
15. Möhler, H. and Okada, T. (1977). Properties of ³H-diazepam binding to benzodiazepine receptors in rat cerebral cortex. *Life Sci.*, **20**, 2101–2110
16. Möhler, H., Okada, T. and Bird, E. (1978). Huntington's chorea: decrease in benzodiazepine-receptor binding. *7th Int. Congress of Pharmacology*, Paris, Abstr. No. 2536
17. Möhler, H., Okada, T., Ulrich, J. and Heitz, Ph. (1978). Biochemical identification of the site of action of benzodiazepines in human brain by ³H-diazepam binding. *Life Sci.*, **22**, 985–996

18. Möhler, H., Polc, P., Cumin, R., Pieri, L. and Kettler, R. (1979). Nicotinamide or brain constituent with benzodiazepine-like action. *Nature*. (In press)
19. Nistri, A. and Constanti, A. (1978). Effects of flurazepam on amino acid evoked responses recorded from the lobster muscle and the frog spinal cord. *Neuropharmacology*, **17**, 127–135
20. Okada, T., Spokes, E. G., Bird, E. D. and Möhler, H. (1979). Huntington's chorea: decrease in benzodiazepine-receptor binding in putamen and caudate nucleus. (In preparation)
21. Okada, T. and Möhler, H. Ligand specificity of the brain benzodiazepine receptor. (In preparation)
22. Pegram, V., Robinson, C., Donaldson, P., Beaton, J. and Smythies, J. (1975). The effects of chronic use of nicotinamide on human sleep. *2nd Int. Congr. on Sleep Research*, Edinburgh, Abstr., p. 91
23. Randall, L. O., Schallek, W., Sternbach, L. H. and Ning, R. Y. (1974). Chemistry and pharmacology of the 1,4-benzodiazepines. In M. Gordon (ed.), *Psychopharmacological Agents*, Vol. III, pp. 175–281
24. Robinson, C. R., Pegram, G. V., Hyde, P. R., Beaton, J. M. and Smythies, J. R. (1977). The effects of nicotinamide upon sleep in humans. *Biol. Psychiatry*, **12**, 139–143
25. Squires, R. F. and Braestrup, C. (1977). Benzodiazepine receptors in rat brain. *Nature*, **266**, 732–734
26. Tallman, J. F., Thomas, J. W. and Gallager, D. W. (1978). GABAergic modulation of benzodiazepine binding site sensitivity. *Nature*, **274**, 383–385
27. Toffano, G., Guidotti, A. and Costa, E. (1978). Purification of an endogenous protein inhibitor for the high-affinity binding of gamma-aminobutyric acid to synaptic membranes of rat brain. *Proc. Natl. Acad. Sci. USA*, **75**, 4024–4028
28. Williamson, M. J., Paul, S. M. and Skolnick, P. (1978). Labelling of benzodiazepine receptors *in vivo*. *Nature*, **275**, 551–553

# 2

## Cerebral Circulation and Metabolism in Sleep

### D. H. Ingvar

### 2.1 INTRODUCTION

The ideal instrument to study the human cerebral circulation and metabolism in sleep – and indeed also in wakefulness – is today in existence. By means of emission tomography, using positron emitting tracers, three-dimensional maps can be obtained of the blood flow and the oxygen uptake, the glucose consumption, etc., in circumscribed parts of the brain[13]. Since this technique is atraumatic, it is eminently suitable for studies in conscious subjects and should also be easily applied to sleep problems. However, the number of functional states so far investigated with emission tomography is very small and no studies of sleep have been done up to now.

The present brief review deals with observations on the cerebral circulation and metabolism in sleep in animals and in man acquired with methods which measure either the *global* cerebral blood flow (CBF) and metabolism or *regional* cerebral blood flow (rCBF). Both approaches are fundamentally based upon tracer techniques devised some 30 years ago. The global technique is the nitrous oxide method of Kety and Schmidt[8] and the regional procedure is the xenon-133 clearance technique to measure rCBF of Lassen and Ingvar[9].

## 2.2 GENERAL RELATIONSHIPS BETWEEN EEG, CEREBRAL BLOOD FLOW AND METABOLISM

It was suggested by Hans Berger that the EEG carries information on the metabolic activity of the neurons[2]. His notion has been supported by much experimental and some clinical studies[5]. Thus, it has been found both in animals and man that the mean frequency of the EEG and the oxygen uptake of the brain correlates[6]. Since cerebral blood flow is controlled by the metabolic activity of the nervous tissue of the brain, it follows that there are also correlations between the EEG and the flow[5]. In general, these correlations imply that the higher the mean frequency is of the EEG (the more desynchronized the pattern is), the higher is the cerebral oxygen uptake and blood flow – and vice versa. There is some indirect evidence that this general principle holds not only globally, for the whole brain, but also for regional EEG changes which, for example, accompany sensory perception, motor performance and mental activity[5].

## 2.3 SLEEP AND THE GLOBAL BRAIN METABOLISM AND BLOOD FLOW

In 1955, Mangold *et al.*[12] performed their classic study with the Kety technique on sleep in six normal subjects. The measurements were made in wakefulness and in slow-wave sleep as monitored by EEG. No significant changes were found, remarkably enough. There was, however, in five of the subjects a small metabolic decrease, but in one a substantial increase was recorded instead.

This finding is at notable variance with the principle mentioned above, that the slow-wave EEG pattern should imply reduction of the cerebral oxygen uptake and blood flow. It thus appears as if normal sleep may constitute a special case in which another type of metabolic and flow regulation is active in the brain than in wakefulness. Here it should be recalled that *the young age* appears to constitute another similar case. Children have a slower EEG than adults, but their cerebral oxygen uptake is nevertheless higher than in adults.

It should here be mentioned that Lübbers *et al.*[11] showed in dogs fluctuating between spontaneous sleep and wakefulness that the sleep phases with slow waves in the EEG were accompanied by a slight but significant decrease of the cortical oxygen uptake.

## 2.4  RESPIRATION AND SYSTEMIC CIRCULATION IN SLEEP

As shown by Bülow[3], normal slow-wave sleep is accompanied by a decreased sensitivity of the respiratory centres to carbon dioxide. This decrease is roughly proportional to the depth of sleep. Slow-wave sleep therefore leads to a moderate hypercapnia of about 5 mmHg and this accounts for a moderate general increase of CBF amounting to about 10–20%. This was noted by Mangold et al.[12] and others.

In REM sleep the carbon dioxide sensitivity increases and attains a level of the same magnitude as in wakefulness[3]. This accounts for the increase and irregularity of respiration which has been observed by many workers during the REM phases. In principle, this hypocapnic tendency brings about cerebral vasoconstriction, but, as will be seen below, the existing evidence indicates that REM sleep – with its increase in neuronal activity – augments both the cerebral oxygen uptake and hence also the blood flow. So in spite of the vasoconstriction there should be a net CBF increase during REM sleep.

Normal sleep, especially during the first part of the night, is also accompanied by a fall of the systemic blood pressure, which often amounts to about 30–50 mmHg (systolic)[13]. This fall of the cerebral perfusion pressure should not affect CBF since there is autoregulation of the cerebral vascular bed, i.e. an adaptation of the width of the vessels to the pressure, so that a constant flow is upheld. However, if the autoregulation is compromised or lost, globally or generally, which may happen following cerebral hypoxia/anoxia, the fall of pressure in sleep might, it seems, be deleterious for the tissue perfusion which then is passively reduced with the pressure. Mechanisms of this type may be operating in elderly subjects with, for instance, so-called multi-infarct dementia who often suffer exacerbation of their symptoms at night with periods of confusion.

In the REM phases there is usually an increase of the blood pressure (and of the pulse rate). These changes should not affect CBF either in the normally autoregulating brain.

## 2.5  REGIONAL CEREBRAL BLOOD FLOW IN SLEEP

With the xenon-133 clearance technique one can measure the rCBF in several brain (cortex) regions simultaneously and obtain a picture of the 'functional landscape' of the hemisphere under study. In resting wakefulness this landscape is typically 'hyperfrontal' with high blood flows in anterior parts of the hemisphere and lower in occipital, parietal and temporal regions. The landscape of resting wakefulness is, in all likelihood, related to a high activity in cortical regions responsible for an inner anticipatory programming of motor behavioural acts, a cerebral activity termed 'simulation of behaviour'[4].

A few accidental observations in our laboratory and by Dr J. Overgaard at Odense University (personal communication) show that the hyperfrontal pattern of wakefulness disappears in sleep (*slow-wave sleep*). Instead a rise in activity has been noted in temporal regions where the activity in resting wakefulness is notably low. Although these observations require confirmation, they indicate that slow-wave sleep may be accompanied by a significant redistribution of rCBF which reflects a redistribution of neuronal activity.

In *REM sleep* there are also changes on the regional plane. Animal experiments have clearly demonstrated that this type of sleep is accompanied by substantial increase of the cerebral blood flow[7] which in certain brain regions may exceed 50%[14]. These flow changes also mirror metabolic increases of corresponding magnitude.

With a relative method to record the regional cerebral blood volume (rCBV) we have succeeded in showing that this parameter (which reflects rCBF alterations) shows regional changes in REM sleep. Increases were seen especially in frontal and parietal areas. This finding suggests two things. First, that the functional cortical landscape in REM sleep is different from that in slow-wave sleep, and, second, that the REM landscape shows certain similarities to those seen in wakefulness in many different forms[4, 15].

## 2.6  CONCLUSION

It should be evident from this review that our present knowledge of the cerebral circulation and metabolism in sleep is very limited, and that there are a number of observations which appear irreconcilable. Notably amongst these is, first, the fact that global measurements have not shown any significant alteration of the cerebral oxygen uptake or blood flow in slow-wave sleep in man[12]. Second, animal studies[14] show clearly that REM sleep should be accompanied by dramatic flow and metabolic increases.

Much of the discrepancy may be dissolved when *regional* methods, preferably based upon emission tomography, are applied systematically in future sleep studies. There is, as mentioned above, reason to believe that both slow-wave and REM sleep are accompanied by significant redistributions of the brain activity (and hence also of the cerebral blood flow). The future study of such redistributions of brain activity will undoubtedly give us much new information of the physiology of sleep in the same manner as such regional studies have given much new information of how the brain functions when awake[10].

## Acknowledgements

Supported by the Swedish Medical Research Council (project no. B78-14X-00084-14C) and by the Wallenberg and Thuring Foundations in Stockholm.

## References

1. Baust, W. and Bohnert, B. (1969). The regulation of heartrate during sleep. *Exp. Brain Res.*, **7**, 169–180
2. Berger, H. (1933). Ueber das Elektroencefalogramm. *Arch. Psychiatr. Nervenkr.*, **101**, 452–458
3. Bülow, K. (1963). Respiration and wakefulness in man. Thesis. *Acta Physiol. Scand.*, **59** (Suppl.), 209
4. Ingvar, D. H. (1975). Patterns of brain activity revealed by measurements of regional blood flow. In D. H. Ingvar and N. A. Lassen (eds.), *Brain Work*, pp. 397–413. (Copenhagen: Munksgaard)
5. Ingvar, D. H., Rosén, I. and Johannesson, G. (1978). EEG related to cerebral metabolism and blood flow. *Contribution to Symposium on Latest Advances in Pharmaco-Encephalography*, July 17–20, Basle. In *Pharmakopsychiat.* (1979). **11**
6. Ingvar, D. H., Sjölund, B. and Ardö, A. (1976). Correlation between dominant EEG frequency, cerebral oxygen uptake and blood flow. *Electroencephalogr. Clin. Neurophysiol.*, **41**, 268–276
7. Kanzow, E. (1965). Changes in blood flow of the cerebral cortex and other vegetative changes during paradoxical sleep periods in the unrestrained cat. In Jouvet (ed.), *Neurophysiologie des Etats de Sommeil*, pp. 231–240, Colloque international CNRS, No. 127, Paris

8. Kety, S. S. and Schmidt, C. F. (1945). The determination of cerebral blood flow in man by the use of nitrous oxide in low concentration. *Am. J. Physiol.*, **143**, 53–66

9. Lassen, N. A. and Ingvar, D. H. (1961). The blood flow of the cerebral cortex determined by radioactive krypton 85. *Experientia (Basel)*, **17**, 42

10. Lassen, N. A., Ingvar, D. H. and Skinhøj, E. (1978). Mapping the functions of the cerebral cortex. *Sci. Am.* October

11. Lübbers, D. W., Ingvar, D. H., Betz, E., Fabel, H., Kessler, M. and Schmahl, F. W. (1964). Sauerstoffverbrauch der Grosshirnrinde in Schlaf- und Wachzustand beim Hund. *Pflügers Arch., Eur. J. Physiol.*, **281**, 58

12. Mangold, R., Sokoloff, L., Conner, E., Kleinerman, J., Therman, P. and Kety, S. S. (1955). The effects of sleep and lack of sleep on the cerebral circulation and metabolism in normal young men. *J. Clin. Invest.*, **34**, 1092–1100

13. Phelps, M. E., Joffman, E. J., Huang, S.-C. and Kuhl, D. (1978). ECAT: A new computerized tomographic imaging system for positron-emitting radiopharmaceutics. *J. Nucl. Med.*, **19**, 635–647

14. Reivich, M., Isaacs, G., Evarts, E. and Kety, S. S. (1968). The effect of slow-wave sleep and REM sleep on regional cerebral blood flow in cats. *J. Neurochem.*, **15**, 301–306

15. Risberg, J. and Ingvar, D. H. (1972). Increase of regional cerebral blood volume during REM-sleep in man. In W. P. Koella and P. Levin (eds.), *Sleep, Physiology, Biochemistry, Psychology, Clinical Implications*. (Basel: Karger)

# 3

# α- and β-Adrenergic Mechanisms in the Control of Sleep Stages

## P. T. S. Putkonen

### 3.1 INTRODUCTION

Norepinephrine (NE) appears intimately connected with central nervous mechanisms controlling various aspects of vigilance[22,25,41], but its explicit role in the mechanisms of paradoxical sleep (PS) has remained elusive and a point of active controversy. To a minor degree the same holds for NE and waking.

Suppression of PS with concurrent fall in cerebral NE following bilateral coagulation of the locus caeruleus (LC)[50] suggested that NE was involved in this state of sleep. The emergent 'noradrenergic hypothesis of PS' was later elaborated to include serotoninergic (5-HT) 'priming', followed by 'triggering'[24] or 'executive'[25] acetylcholine – (ACh) and noradrenergic mechanisms. Thus, according to Jouvet[25], the three neurotransmitters or modulators would, in succession or co-operation, all contribute to the realization of PS. Confluent neuropharmacological data from several experimental lines, however, has led others[17,55] to opposite conclusions postulating a negative correlation between synaptic availability of NE and the amount of PS[17,55]. Among the strongest arguments for the latter hypothesis has been the finding that partial depletion of NE following inhibition of its synthesis by α-methylparatyrosine (αMPT)

19

increases PS in rats[17] and cats[28]. Some earlier discrepancies may be accounted for by toxic or irritating effects of the drug, but the recent results of Kafi *et al.*[26] cast serious doubts on conclusions drawn from the cited experiments. Serial injections of $a$MPT methylester circumvent the problem of renal toxicity and permit greater depletion of catecholamines than reached by single injections. In the rat repeated injections of $a$MPT produced profound depletion of cerebral catecholamines paralleled, after transient enhancement, by a progressive decrease in PS[26].

On the other hand, a recent study[23] re-examining the effects of massive LC lesions in the cat reported recovery to normal amounts of PS, with some permanent qualitative changes, within the second postoperative week in spite of persisting 60–85% reduction of cerebral NE. It must be concluded that the problematic relations of NE neurons and PS remain largely unsettled.

In the waking state, convergent lines of evidence relate catecholamines to locomotor and EEG activation[13,22,25,51], but in fact, even at these points, pharmacological literature is far from unanimous[7,34]. Furthermore, early quantitative normalization of EEG arousal and normal amphetamine response after massive LC lesions in the cited experiments by Jones *et al.*[23], despite a mean 85% decrement of NE at the telencephalic level, led the authors (the first of them among cited proponents[22] of the noradrenergic waking theory) to 'question the very basis of the pharmacological model of involvement of NE neurons in processes of waking and arousal'.

For the past 3 years my laboratory has been working on the problem of noradrenergic mechanisms and the control of the stages of vigilance. We have focused on the $a$- and $\beta$-adrenergic receptors by the use of various adrenergic agonists and antagonists. A short synopsis of our results on $a$-receptors and sleep[45] was presented at the IUPHAR satellite symposium 'Pharmacology of the States of Alertness' in Montpellier, July 1978.

## 3.2 METHODS

The results are based on 16 h undisturbed recordings from 35 cats implanted with fixed electrodes for EEG (sensorimotor–visual cortex), PGO 'spikes' (lateral geniculate nucleus), EOG (supraorbital bone) and EMG (neck muscles). The records were visually classified into five stages of vigilance in close accordance with conventional criteria[54,59]. The waking state (W) consists of the aroused (A) and drowsy (D) stages. The former is characterized by low voltage, mixed frequency EEG, tonic EMG, often with abundant movement artefacts and eye movements. Stage D is a relatively quiescent stage with long trains of rather high, synchronized 3–8 Hz waves in the EEG. Slow-wave sleep (SWS) is divided into a light stage (S1) with spindles and bursts of irregular slow waves and a deep stage (S2) with high, poorly synchronized slow activity exceeding 50% of the epoch (1 min). Paradoxical sleep (PS) is characterized by low fast EEG, minimal or absent EMG, PGO spikes and rapid eye movements.

After visual analysis the coded data were fed to a computer used for statistical analysis and production of hypnograms (Figures 3.1 and 3.4).

Control recording after intraperitoneal saline injections were obtained from each animal prior to drug experiments. All the drugs were also given intraperitoneally. In experiments combining an agonist with an antagonist, the latter was always given 10 min before the agonist. The last injection was at the start of the evaluated 16 h period fixed between 3.30 p.m. and 7.30 a.m. ($\pm$ 30 min). The same cats served in several experiments, with a rest and wash-out period of at least 1 week separating different drug recordings.

## 3.3 RESULTS AND COMMENTS

### 3.3.1 α-Adrenergic agonists

Two recognized α-adrenergic stimulants, clonidine (Clo), 5, 10 and 20 μg/kg, and xylazine (Xyl), 0.5, 1 and 2 mg/kg, were found to produce a dose-dependent suppression of PS with a parallel increase in drowsy waking (D)[32,45]. Both drugs also markedly decreased deep slow-wave sleep (S2); see Figure 3.1.

α-Methyldopa (αMD) is also considered as an indirect α-agonist because its metabolites, αMNE and αMDA, are potent α-receptor stimulants[30], and do not qualify as 'false transmitters' interfering passively with noradrenergic transmission. The latter concept has been used to explain αMD's amply-documented suppressive effect

21

on PS, and as an argument for the noradrenergic hypothesis[25]. In our hands[31,44] 100 mg/kg of aMD decreased the 16 h mean percentage of PS to 1.2%, which was comparable to the effect of 2 mg/kg of Xyl. The biggest dose of Clo used in our studies did not fully match these effects (3.0% PS in 16 h), but the minimal 20 μg/kg dose required for this makes Clo by far the most potent of the three agonists. aMD differed from the two direct agonists, by a relatively short inhibitory effect on S2, followed by a trend to increased S2 (Figure 3.1), making up for the initial decrement in the 16 h percentage, which reached control levels[31].

CONTROL

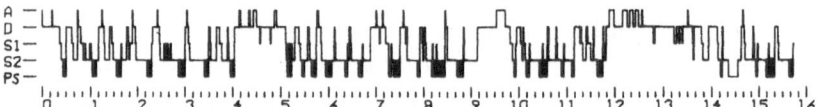

CLONIDINE 10 μG/KG

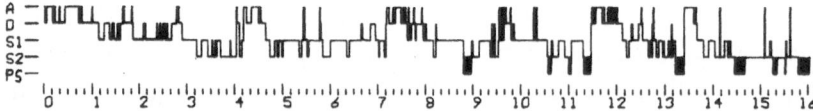

XYLAZINE 1 MG/KG

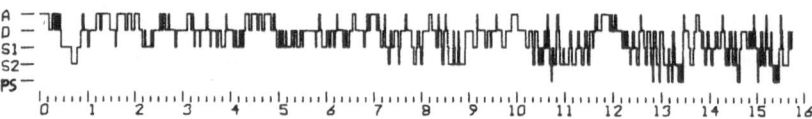

ALPHA-METHYLDOPA 100 MG/KG

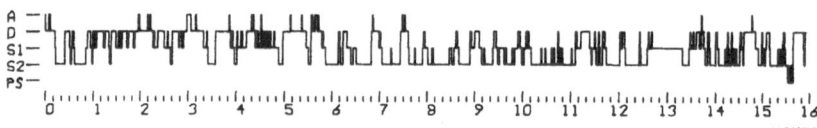

HOURS

**Figure 3.1** Representative 16 h hypnograms after i.p. injection of saline ( = control) and after injections of three a-agonists. A = aroused, D = drowsy, S1 and S2 = light and deep slow-wave sleep, PS = paradoxical sleep

Our results with Clo appear to be in fair agreement with Klein-logel *et al.*[29], who first reported suppression of PS, decrease of SWS and increased 'dozing' after 200 μg/kg of Clo in rats. Decreased PS after Clo (300 μg *per os*) has also been described in humans[1].

On the other hand, our results are not in accord with several

22

earlier studies assimilating behavioural sedation or sleep-like states, and EEG synchrony or slowing after Clo[8, 20] and other α-agonists[7] with physiological sleep. In the cat both Clo and Xyl inhibit deep SWS and increase drowsy waking and even the stupor-like sedation with hypersynchronous EEG produced by anaesthetic doses of Xyl fits neither the behavioural nor EEG criteria of physiological sleep. In rats, 200 μg/kg of Clo increased 'dozing' and decreased deep SWS[29]. In man, a relatively small dose (300 μg/subject) of Clo increased mainly stages 1 and 2 (not given separately)[1]. Stages 3 and 4 are not included in the data, but it may be inferred from a table that 3 and 4 remained at control level. While it is not possible to sort out differences in classification and dosage from possible differences between the three species studied, it seems likely that Clo consistently disturbs rather than promotes physiological sleep.

### 3.3.2 α-Adrenergic antagonists

Among five different α-adrenergic receptor blocking drugs studied, phentolamine (Phe) had an exceptional capacity to increase PS[42, 53]. With 20 mg/kg the 16 h mean of PS increased from 14 to almost 23% of recording time ($p < 0.001$). This is a 62% increment, but during the maximal effect between 4 and 8 h after injection, PS peaked to 120% above baseline[42]. In another group of cats Phe increased the rebound following 72 h of PS-deprivation from 24% PS/24 h in non-drugged rebound sleep to 37% when 20 mg/kg was injected at the beginning of the rebound[47]. The increase in PS after Phe is mainly due to lengthening of the periods. This effect becomes more exaggerated after PS-deprivation, which even alone increases the duration but also the number of PS periods (Figure 3.2). In kittens, at the age of 2 weeks, 20 mg/kg of Phe increased PS from 46 to 68% in 8 h recording and the mean duration of PS periods from 5 to 8 min[53].

None of the other α-blockers, tolazoline (Tol), yohimbine (Yoh), thymoxamine (Thy) or phenoxybenzamine (Pbz), significantly increased PS in spite of trials at several doses, especially with Thy (5, 10, 10 mg/kg), which in humans (150 mg i.v./subject) was reported to produce a transient increase in early night REM sleep[39], and Pbz (5, 10 and 20 mg/kg) which in rats (40 mg/kg *per os*) increased 24 h mean of PS 28% above control[16]. In our series there was a trend towards decreased PS after Pbz, in line with Matsumoto and Watanabe[35], who report decreased PS in two cats receiving 15 mg/kg of Pbz.

Many of the $\alpha$-blockers increased waking in the early part of the record, as was also the case with Thy in humans[39]. The arousing effect was most pronounced after Yoh (2 mg/kg), which in the first

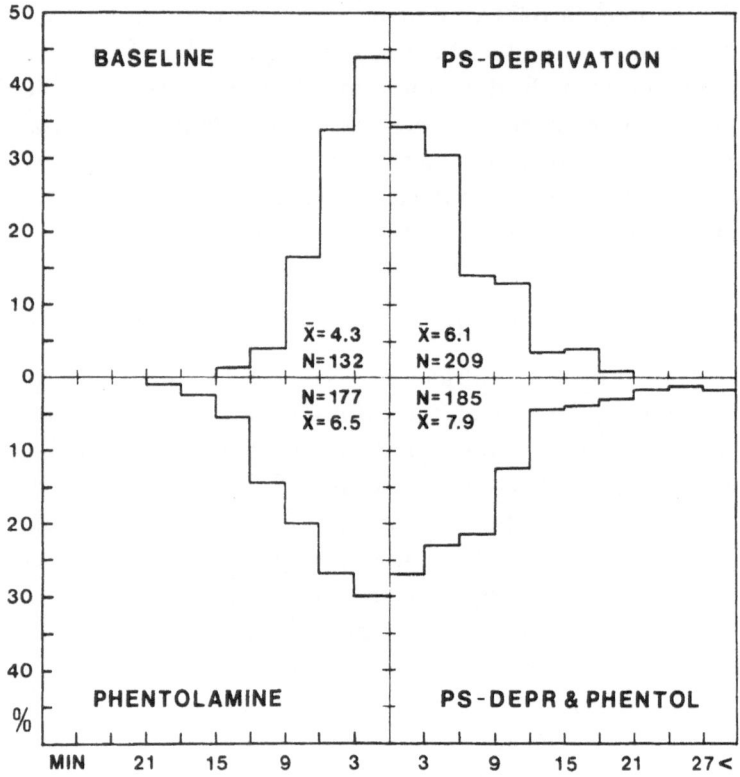

**Figure 3.2** Relative distribution of PS periods of different duration in 12 h of sleep in standard laboratory conditions and following 72 h of PS deprivation with or without phentolamine (20 mg/kg i.p.). All data from the same group of six cats

4 h increased waking from a control value of 31% to 71% ($p < 0.01$)[46]. Sleep gradually normalized within the next 4 h. Such arousing effects, previously reported with piperoxane (5 mg/kg) in the rat[13], could, in some $\alpha$-blockers, mask a potential PS enhancing property, manifest only at a critically narrow dose level, and sensitive even to individual variations.

### 3.3.3  β-Adrenergic antagonists

Two β-blockers, pindolol (0.1 and 0.5 mg/kg) and propranolol (5 mg/kg), decreased deep SWS, with a corresponding increase in drowsy waking[18]. The effects were clearest for propranolol (Pro), which also significantly decreased PS (Figure 3.3).

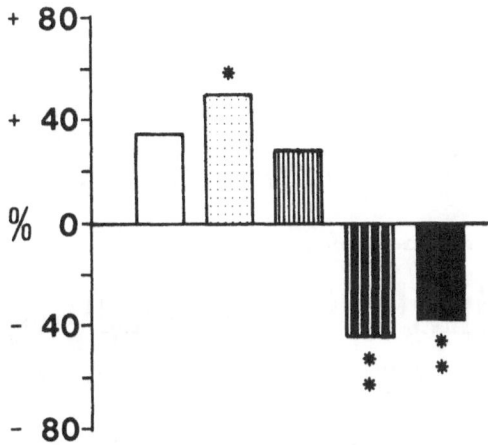

**Figure 3.3**  Relative changes of 16 h means ($n = 7$) of stages of vigilance in per cent of respective controls after propranolol (5 mg/kg i.p.). From left to right, in increasingly dense shading: A, D, S1, S2 and PS. There is a significant ($p < 0.01$) decrease in S2 and PS and increase ($p < 0.05$) in D after Pro

These results are at variance with those of Hartmann and Zwilling[16], who found no changes in sleep after 3–40 mg/kg of propranolol in rats. Sleep laboratory findings in normal humans are either negative (no change with 120 mg propranolol on seven consecutive nights)[12] or transient (decrease in sleep efficiency index after 400 mg of metoprolol and moderate increase in percentage REM after 200 mg of acebutolol, evident only on the first night on drugs)[27]. This is rather intriguing in view of recurrent complaints of sleep disturbances (insomnia or bizarre dreaming) in patients treated with different β-blockers, and considering the wide distribution of β-receptors in the brain (see reference 18). Contrary to the α-receptive systems, β-receptors do not appear to have an important part in the direct control of the normal sleep–waking cycle, but functions mediated by them may become critical in the presence of some latent fragility in the regulation of sleep.

### 3.3.4 Agonist–antagonist interactions

To test whether the sleep changes produced by the α-agonists are indeed due to their α-adrenoceptor stimulating property, the effects of Clo were studied after pretreatment with the five α-blockers mentioned above and after β-blockade with propranolol (Figures 3.4 and 3.5).

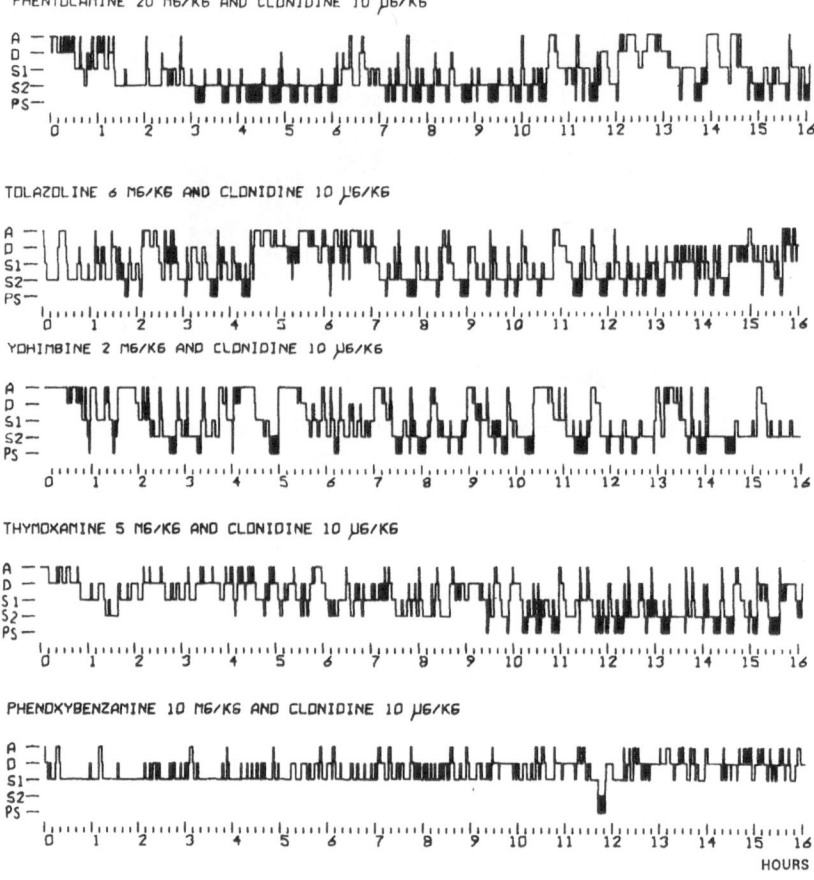

**Figure 3.4** Hypnograms showing interactions of five α-antagonists with a standard dose of clonidine. Compare with Figures 3.1 and 3.5

The PS suppressing effects of a standard dose (10 μg/kg) of Clo were completely overridden by Phe (10 and 20 mg/kg), which even with Clo increased PS significantly above control[45]. Tol and Yoh also clearly antagonized the PS inhibiting effect of Clo. Thy was ineffective and Pbz (10 mg/kg) potentiated Clo's effects on both PS and S2, leading to nearly total suppression of these stages in the 16 h records (PS 1.8%, S2 2.3%). There was a corresponding increase in D with little change in A or S1 (Figure 3.4).

β-Blockade with 5 mg/kg of Pro did not antagonize, but potentiated the detrimental effects of Clo on both PS and S2 (Figure 3.5). The effects were similar but less profound than after Pbz.

Furthermore, Phe was also shown to antagonize the PS and S2 inhibiting effects of Xyl[45] and the PS suppressing effect of αMD[31,44], which indicates that these drugs also inhibit PS through an α-agonist mechanism.

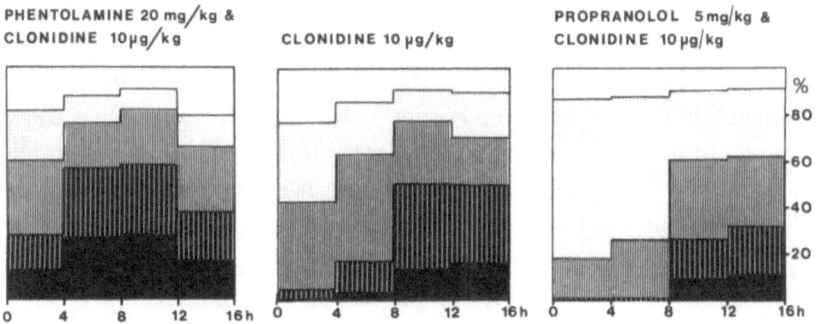

**Figure 3.5** Sequential 4 h mean percentages of the stages in cats receiving Clo as such or after pretreatment with the α-blocker Phe or the β-blocker Pro. The exaggerated antagonism by Phe is contrasted by the synergistic effects of Pro, leading to total suppression of PS (black) and to almost total suppression of S2 (heavy stripes) during the first 8 h after injection of Clo. From top to bottom A to PS (in increasing shading as in Figure 3.3)

Delbarre and Schmitt[7,8] showed that behavioural sedation or 'sleep' caused by Clo and Xyl in chicks and mice could be antagonized with Tol, Phe, Yoh, piperoxane and dibenamine, but not by Pbz nor by any of the several β-blockers investigated. This led them to postulate the existence of central nervous α-adrenergic receptors with a sedative function and pharmacological properties different from the corresponding peripheral receptors. The agonist–antagonist interactions with regard to PS control in cats fit into the above pharmacological pattern and suggest that they are dependent on the same kind of receptors.

## 3.4  GENERAL DISCUSSION AND CONCLUSIONS

### 3.4.1  Subdivision of $\alpha$-adrenergic receptors

During recent years a rapid accumulation of knowledge on the nature of the $\alpha$-adrenoceptors has led to their division into two main types[3,52], existing both in the periphery[61] and in the central nervous system[36,58]. These are: (1) the classic post-synaptic or $\alpha_1$-type receptors mediating e.g. vasoconstriction in the periphery and excitatory responses in cerebral cortical cells[4]; and (2) the pre-synaptic or $\alpha_2$-type receptors, best known as autoreceptors mediating negative feedback inhibition of NE release and synthesis[15,52,57]. Type $\alpha_2$-receptors also inhibit cholinergic nerves in the guinea pig ileum[61] and receptors inhibiting ACh release in the cerebral cortex[2] could be of the $\alpha_2$ variety.

Clo, Xyl and $\alpha$MNE all have selective preference for $\alpha_2$-receptors[52,61]. Among the effective antagonists, Yoh and Tol have preference for pre-synaptic receptors, whereas Thy and Pbz show selectivity to post-synaptic $\alpha_1$-adrenoceptors[10,11,15]. The relatively short-acting competitive $\alpha_1$-blocker Thy did not significantly affect PS suppression after Clo, whereas Pbz, inducing a long-lasting non-equilibrium blockade of mainly post-synaptic $\alpha_1$-receptors dramatically aggravated Clo's effects on sleep. Phe has recently been regarded as a preferentially pre-synaptic $\alpha$-blocker[10,11], but it also has considerable affinity to $\alpha_1$-receptors[58].

### 3.4.2  Paradoxical sleep

Our results imply that stimulation of $\alpha_2$-adrenoceptors inhibits PS and that this effect can be overcome by several $\alpha_2$- but not by $\alpha_1$- nor $\beta$-receptor blocking drugs.

There is compelling evidence for the existence of $\alpha_2$-type inhibitory autoreceptors especially sensitive to Clo in noradrenergic neurons of the LC[5,56]. Antagonism of noradrenergic inhibition of stimulation-induced NE release by Phe from brain slices[57] indicates that inhibitory pre-synaptic $\alpha_2$-autoreceptors may also be found at the cortical level. Activation of $\alpha_2$-type receptors on the soma or dendrites of LC neurons inhibits their firing[56], and the release of NE may further be decreased by prejunctional inhibitory autoreceptors in cortical fibres. Thus, our main result, suppression of PS with $\alpha_2$-type agonists, antagonized by preferential $\alpha_2$- but not $\alpha_1$- nor $\beta$-

blockers, is consistent with the hypothesis that NE is an essential factor in the successful elaboration of PS. Similar conclusions have recently been drawn by Autret *et al.*[1] in their study demonstrating that Yoh antagonizes REM sleep inhibition by Clo in humans, by Gaillard and Kafi[14] discussing the antagonism of small doses of chlorpromazine and Clo on PS in rats and humans, and by Depoortere[9] on the basis of his PGO experiments.

Granting a positive role for NE in the mechanisms of PS, the optimal level of NE release for this stage may be less than during active waking. PS periods are often terminated by awakening, which could be precipitated by progressive build-up of noradrenergic transmission to a critical level. Moderate inhibition of noradrenergic transmission might thus stabilize and prolong PS, which could explain why inhibition of NE synthesis initially increases PS[17, 26. 28]. Furthermore, a fortuitous balance between the $\alpha_1$- and $\alpha_2$-blocking potencies could be a factor in the exceptional ability of Phe to increase PS. Electrically-induced release of NE is maximally facilitated or inhibited by pre-synaptic α-antagonists or α-agonists with low ( $\leqslant 1$ Hz) pulse frequencies[52]. Consequently, such drugs may critically affect NE neurons at the shift from SWS to PS and during PS when a slow[19, 33] or intermittent[6] activity prevails in LC and sub-coeruleus neurons.

The desynchronized EEG during PS and active waking is associated with increased cortical ACh liberation[21]. α-Receptor mediated inhibition of ACh release[2] could thus contribute to the PS suppressing and sedative effects of $\alpha_2$-agonists. As this system also is most effective at low impulse frequencies it may hinder entrance into PS from SWS and yet permit arousal by stimuli leading to sufficiently intense firing in ACh neurons.

Intricate interaction between the noradrenergic and ACh-ergic systems is indicated by our observation that the PS enhancing effect of Phe can be turned to suppression below control levels in combination with 75 µg/kg of the muscarinic blocker atropine, which alone was an insufficient dose to affect sleep parameters in the cat[43].

### 3.4.3 Slow-wave sleep

The inhibitory effects of Clo and Xyl on S2 are difficult to understand along adrenergic lines, but may depend on inhibition of the

5-HT system. Clo has been demonstrated to inhibit firing of raphe neurons[56] and 5-HT turnover[49]. The synergistic effects of Pbz and Pro with Clo in suppressing S2 could largely or partially depend on the 5-HT receptor blocking property of these drugs[37,38,60]. Since some of our effective $a_2$-antagonists also have 5-HT blocking side-effects[38], but antagonized rather than potentiated Clo's effect on S2[46], blockade of the $a_2$-type receptors seems to be a crucial factor. One could postulate inhibitory $a_2$-receptors on 5-HT neurons, which would fit the notion of inhibitory interrelations between the 5-HT and noradrenergic systems proposed by Jouvet. Iontophoretic recordings by Svensson et al.[56], however, speak against the existence of inhibitory $a$-receptors in the dorsal raphe nucleus. These authors tend to regard the inhibitory effect of systemically administered Clo on raphe neurons as an indirect consequence of impaired NE transmission.

### 3.4.4 Waking

The disclosure of two functionally different classes of $a$-receptors may bring light to the controversial literature regarding the arousing versus sedative effects of catecholamines.

The selective $a_1$-agonist, methoxamine, increases locomotor activity in rats[51] and restlessness with EEG arousal at the expense of SWS and PS in dogs[40]. The sleep changes could be antagonized by Pbz, which in this case significantly increased SWS ($\sim$ S2) and to some extent even PS. This is directly contrary to the effects of Pbz when combined with the selective $a_2$-agonist Clo in our experiments, but fits the notion of the predominantly excitatory effects of cortical $a$-receptors[4], which may be inferred to be of the $a_1$ variety.

Selective $a_2$-agonists, on the other hand, may produce their sedative and synchronizing effects[7,8,29] through autoreceptor mediated inhibition of firing[5] and transmitter release[57] in NE neurons, combined with reduced cortical ACh output due to activation of inhibitory pre-synaptic $a$-receptors on ACh-ergic endings[2]. Conversely, the arousing effects of potent $a_2$-blockers like Yoh and piperoxane[13] could result from increased NE and ACh output, liberated from inhibitory noradrenergic and partly adrenergic influences engaged in the physiological control of impulse activity and/or transmitter release[5,52].

## Acknowledgements

This review is based on experiments performed in cooperation with I. Hilakivi, T. Kovala, A. Leppävuori, J. Mäkelä and D. Stenberg.

Our work has been supported by grants from the Medical Research Council of the Academy of Finland and from the Research Foundation of Orion Corporation Ltd.

## References

1. Autret, A., Mintz, M., Beillevaire, T., Cathala, H.-P. and Schmitt, H. (1977). Effect of clonidine on sleep patterns in man. *Eur. J. Clin. Pharmacol.*, **12**, 319

2. Beani, L., Bianchi, C., Giacomelli, A. and Tamberi, F. (1978). Noradrenaline inhibition of acetylcholine release from guinea-pig brain. *Eur. J. Pharmacol.*, **48**, 179

3. Berthelsen, S. and Pettinger, W. A. (1977). A functional basis for classification of α-adrenergic receptors. *Life Sci.*, **21**, 595

4. Bevan, P., Bradshaw, C. M. and Szabadi, E. (1977). The pharmacology of adrenergic neuronal responses in the cerebral cortex: evidence for excitatory α- and inhibitory β-receptors. *Br. J. Pharmacol.*, **59**, 635

5. Cedarbaum, J. M. and Aghajanian, G. K. (1977). Catecholamine receptors on locus coeruleus neurons: pharmacological characterization. *Eur. J. Pharmacol.*, **44**, 375

6. Chu, N. and Bloom, F. E. (1972). Norepinephrine-containing neurons: changes in spontaneous discharge patterns during sleeping and waking. *Science*, **179**, 908

7. Delbarre, B. and Schmitt, H. (1971). Sedative effects of α-sympathomimetic drugs and their antagonism by adrenergic and cholinergic blocking drugs. *Eur. J. Pharmacol.*, **13**, 356

8. Delbarre, B. and Schmitt, H. (1973). A further attempt to characterize sedative receptors activated by clonidine in chickens and mice. *Eur. J. Pharmacol.*, **22**, 355

9. Depoortere, H. (1978). Effects of some pre- and post-synaptic noradrenergic antagonists on the central action of clonidine. Presented at the *18th Annual Meeting APSS*, April 5–9, Palo Alto

10. Doxey, J. C., Smith, C. F. C. and Walker, J. M. (1977). Selectivity of blocking agents for pre- and postsynaptic α-adrenoceptors. *Br. J. Pharmacol.*, **60**, 91

11. Drew, G. M. (1977). Pharmacological characterization of the presynaptic α-adrenoceptor in rat vas deferens. *Eur. J. Pharmacol.*, **42**, 123

12. Dunleavy, D. L. F., MacLean, A. W. and Oswald, I. (1971). Debrisoquine, guanethidine, propranolol and human sleep. *Psychopharmacologia (Berlin)*, **21**, 101

13. Fuxe, K., Lidbrink, P., Hökfelt, T., Bolme, P. and Goldstein, M. (1974). Effects of piperoxane on sleep and waking in the rat: evidence for increased waking by blocking inhibitory adrenaline receptors on the locus coeruleus. *Acta Physiol. Scand.*, **91**, 566

14. Gaillard, J.-M. and Kafi, S. (1978). Brain catecholaminergic control of paradoxical sleep: pharmacological studies. In P. Passouant (ed.), *Pharmacology of the States of Alertness*. (New York: Pergamon Press) (In press)

31

15. Grabowska, M. and Andén, N.-E. (1976). Noradrenaline synthesis and utilization: control by nerve impulse flow under normal conditions and after treatment with alpha-adrenoreceptor blocking agents. *Naunyn-Schmiedeberg's Arch. Pharmacol.*, **292**, 53

16. Hartmann, E. and Zwilling, G. (1976). The effect of alpha and beta adrenergic receptor blockers on sleep in the rat. *Pharmacol. Biochem. Behav.*, **5**, 135

17. Hartmann, E., Bridwell, T. J. and Schildkraut, J. J. (1971). Alpha-methylparatyrosine and sleep in the rat. *Psychopharmacologia (Berlin)*, **21**, 157

18. Hilakivi, I., Mäkelä, J., Leppävuori, A. and Putkonen, P. T. S. (1978). Effects of two adrenergic β-receptor blockers on the sleep cycle of the cat. *Med. Biol.*, **56**, 138

19. Hobson, J. A., McCarley, R. W. and Wyzinski, P. W. (1974). Sleep cycle oscillation: reciprocal discharge by two brainstem neuronal groups. *Science*, **189**, 55

20. Holman, R. B., Shillito, E. E. and Vogt, M. (1971). Sleep produced by clonidine (2-(2,6-dichlorophenylamino)-2-imidazoline hydrochloride). *Br. J. Pharmacol.*, **43**, 685

21. Jasper, H. H. and Tessier, J. (1970). Acetylcholine liberation from cerebral cortex during paradoxical (REM) sleep. *Science*, **172**, 601

22. Jones, B. E. (1972). The respective involvement of noradrenaline and its deaminated metabolites in waking and paradoxical sleep: a neuropharmacological model. *Brain Res.*, **39**, 121

23. Jones, B. E., Harper, S. T. and Halaris, A. E. (1977). Effects of locus coeruleus lesions upon cerebral monoamine content, sleep–wakefulness states and the response to amphetamine in the cat. *Brain Res.*, **124**, 473

24. Jouvet, M. (1969). Biogenic amines and the states of sleep. *Science*, **163**, 32

25. Jouvet, M. (1972). The role of monoamines and acetylcholine-containing neurons in the regulation of the sleep–waking cycle. *Ergebn. Physiol.*, **64**, 166

26. Kafi, S., Bouras, C., Constantinidis, J. and Gaillard, J.-M. (1977). Paradoxical sleep and brain catecholamines in the rat after single and repeated administration of alpha-methylparatyrosine. *Brain Res.*, **135**, 123

27. Kayed, K. and Godtlibsen, O. B. (1977). Effects of the β-adrenoceptor antagonists acebutolol and metoprolol on sleep pattern in normal subjects. *Eur. J. Clin. Pharmacol.*, **12**, 323

28. King, C. D. and Jewett, R. E. (1971). The effects of α-methyltyrosine on sleep and brain norepinephrine in cats. *J. Pharmacol. Exp. Ther.*, **177**, 188

29. Kleinlogel, H., Scholtysik, G. and Sayers, A. C. (1975). Effects of clonidine and BS 100-141 on the EEG sleep pattern in rats. *Eur. J. Pharmacol.*, **33**, 159

30. Kobinger, W. (1978). Central α-adrenergic systems as targets for hypotensive drugs. *Rev. Physiol. Biochem. Pharmacol.*, **81**, 39

31. Leppävuori, A. and Putkonen, P. T. S. (1978). Evidence for central alpha adrenoceptor stimulation as the basis of paradoxical sleep suppression by alpha methyldopa. *Neurosci. Lett.*, **9**, 37

32. Leppävuori, A., Putkonen, P. T. S. and Stenberg, D. (1976). Sedation and suppression of paradoxical sleep in the cat after clonidine. *Acta Physiol. Scand.*, (Suppl. 440), 60

33. McGinty, D. J., Harper, R. M. and Fairbanks, M. K. (1974). Neuronal unit activity and the control of sleep states. In E. D. Weitzmann (ed.), *Advances in Sleep Research*, Vol. I, pp. 173–216. (New York: Spectrum)

34. Marley, E. (1966). Behavioural and electrophysiological effects of catechol-amines. *Pharmacol.Rev.*, **18**, 753
35. Matsumoto, J. and Watanabe, S. (1967). Paradoxical sleep: effects of adrenergic blocking agents. *Proc. Japan Acad.*, **43**, 680
36. Miach, P. J., Dausse, J.-P. and Meyer, P. (1978). Direct biochemical demonstration of two types of α-adrenoreceptor in rat brain. *Nature (London)*, **274**, 492
37. Middlemiss, D. N., Blakeborough, L. and Leather, S. R. (1977). Direct evidence for an interaction of β-adrenergic blockers with the 5-HT receptor. *Nature (London)*, **267**, 289
38. Nickerson, M. and Collier, B. (1975). Drugs inhibiting adrenergic nerves and structures innervated by them. In L. S. Goodman and A. Gilman (eds.), *The Pharmacological Basis of Therapeutics*, 5th Ed., pp. 533–564. (New York: Macmillan)
39. Oswald, I., Thacore, V. R., Adam, K., Březinová, V. and Burack, R. (1975). α-Adrenergic receptor blockade increases human REM sleep. *Br. J. Pharmacol.*, **2**, 107
40. Pickworth, W. B., Sharpe, L. G., Nozaki, M. and Martin, W. R. (1977). Sleep suppression induced by intravenous and intraventricular infusions of methoxamine in the dog. *Exp. Neurol.*, **57**, 999
41. Putkonen, P. T. S. (1974). Monoamines and the regulation of vigilance and sleep. *Med. Biol.*, **52**, 193
42. Putkonen, P. T. S. and Leppävuori, A. (1977). Increase in paradoxical sleep in the cat after phentolamine, an alpha-adrenoceptor antagonist. *Acta Physiol. Scand.*, **100**, 488
43. Putkonen, P. T. S. and Leppävuori, A. (1978). Interactions of central cholinergic and alpha-adrenergic mechanisms in the control of paradoxical sleep. *Neurosci. Lett.*, (Suppl. 1), S336
44. Putkonen, P., Leppävuori, A., Hilakivi, I., Mäkelä, J. and Stenberg, D. (1977). Paradoxical sleep inhibition by α-methyldopa due to central α-adrenoceptor stimulation. *Electroencephalogr. Clin. Neurophysiol.*, **43**, 525
45. Putkonen, P. T. S., Leppävuori, A., Hilakivi, I. and Mäkelä, J. (1978). Central alpha adrenoceptors and the states of vigilance: neuropharmacological experiments in cats. In P. Passouant (ed.), *Pharmacology of the States of Alertness.* (New York: Pergamon Press) (In press)
46. Putkonen, P. T. S., Leppävuori, A. and Stenberg, D. (1977). Paradoxical sleep inhibition by central alpha-adrenoceptor stimulant clonidine antagonized by alpha-receptor blocker yohimbine. *Life Sci.*, **21**, 1059
47. Putkonen, P., Leppävuori, A., Stenberg, D. and Hilakivi, I. (1977). The effect of α-adrenoceptive blockade on the rebound of paradoxical sleep following its selective deprivation. In *Fifth Scandinavian Meeting on Physiology and Behaviour*, p. 35, Helsinki
48. Putkonen, P. T. S., Mäkelä, J., Hilakivi, I., Kovala, T. and Leppävuori, A. (1978). Interactions of propranolol with two drugs affecting monoamines and sleep in the cat. In *Fourth European Congress on Sleep Research*, p. 126. (Tirgu-Mures: Tipografia Tg-Mures, 117/1978)
49. Rochette, L. and Bralet, J. (1975). Effect of the norepinephrine receptor stimulating agent 'clonidine' on the turnover of 5-hydroxytryptamine in some areas of the rat brain. *J. Neural Transm.*, **37**, 19
50. Roussel, B., Buguet, A., Bobillier, P. and Jouvet, M. (1967). Locus coeruleus, sommeil paradoxal, et noradrénaline cérébrale. *C. R. Soc. Biol. (Paris)*, **161**, 2537

51. Segal, D. S. and Geyer, M. A. (1976). Pre- and post-junctional super-sensitivity: differentiation by intraventricular infusions of norepinephrine and methoxamine. *Psychopharmacology*, **50**, 145

52. Starke, K. (1977). Regulation of noradrenaline release by pre-synaptic receptor systems. *Rev. Physiol. Biochem. Pharmacol.*, **77**, 1

53. Stenberg, D., Miettinen, M., Miettinen, K. and Putkonen, P. T. S. (1978). Effect of alpha-adrenoceptor blockade on sleep in kitten. *Neurosci. Lett.*, (Suppl. 1), S337

54. Sterman, M. B., Knauss, T., Lehmann, D. and Clemente, C. D. (1965). Circadian sleep and waking patterns in the laboratory cat. *Electroencephalogr. Clin. Neurophysiol.*, **19**, 509

55. Stern, W. C. and Morgane, P. J. (1974). Theoretical view of REM sleep function: maintenance of catecholamine systems in the central nervous system. *Behav. Biol.*, **11**, 1

56. Svensson, T. H., Bunney, B. S. and Aghajanian, G. K. (1975). Inhibition of both noradrenergic and serotonergic neurons in brain by the α-adrenergic agonist clonidine. *Brain Res.*, **92**, 291

57. Taube, H. D., Starke, K. and Borowski, E. (1977). Presynaptic receptor systems on the noradrenergic neurones of rat brain. *Naunyn-Schmiedeberg's Arch. Pharmacol.*, **299**, 123

58. U'Prichard, D. C., Charness, M. E., Robertson, D. and Snyder, S. H. (1978). Prazosin: differential affinities for two populations of α-noradrenergic receptor binding sites. *Eur. J. Pharmacol.*, **50**, 87

59. Ursin, R. (1968). The two stages of slow wave sleep in the cat and their relation to REM sleep. *Brain Res.*, **11**, 347

60. Weinstock, M., Weiss, C. and Gitter, S. (1977). Blockade of 5-hydroxytrypt-amine receptors in the central nervous system by β-adrenoceptor antagonists. *Neuropharmacology*, **16**, 273

61. Wikberg, J. E. S. (1978). Pharmacological classification of adrenergic α receptors in the guinea pig. *Nature (London)*, **273**, 164

# 4

# Brain Catecholaminergic Activity in Relation to Sleep

## J.-M. Gaillard

Numerous experimental results, obtained in the last 10 years, have indicated that brain catecholaminergic systems are involved in the regulation of wakefulness and sleep. The investigation of the wakefulness mechanisms has given rather unambiguous results, pointing to a role of ascending neuronal systems containing norepinephrine (NE). Selective lesions of these systems by microinjection of 6-hydroxydopamine (6-OHDA), a drug inducing neurotoxic lesions of catecholamine (CA) neurons, result in a decrease of EEG waking, most pronounced during the first 4 days after the lesion, and followed by a gradual recovery[11]. Similarly, the administration of alpha-methylparatyrosine ($a$MPT), an inhibitor of CA synthesis through blockade of tyrosine hydroxylase, induces in animals a decrease of wakefulness, which can be temporarily reversed by microinjections of NE in the preoptic area or in the reticular activating systems[19]. These and other results indicate that an enhancement of activity in NE synapses induces EEG waking, whereas dopaminergic (DA) systems may be more specifically involved in behavioural wakefulness[3].

The role of brain CA systems in the regulation of paradoxical sleep (PS) is much more controversial. Matsumoto and Jouvet[13] reported that reserpine suppresses PS in the cat for several days, and that administration of dopa, the immediate precursor of DA and

35

NE, can transiently make PS reappear. It was shown later that there is no good correlation between the level of endogenous CA and the time course of PS suppression after reserpine[17]. On the third day after intraperitonal administration of reserpine to cats, PS returned to a normal level whereas brain level of CA was still depressed by about 50% with respect to controls.

The effect of chemical or surgical lesions on NE ascending systems also gave variable results. After intracisternal injection of 500 $\mu$g of 6-OHDA, Hartmann et al.[2] described a decrease of waking and a small enhancement of PS. In contrast, Laguzzi et al.[9] reported a dose-dependent decrease of PS after intracisternal administration of 300 $\mu$g to 5 mg of 6-OHDA. Similarly in the rat, 500 $\mu$g of 6-OHDA resulted in a depression of PS during the first 5 days, with a gradual return to normal after about 6 days[12]. As pointed out by Pujol et al.[14], the interpretation of data obtained after 6-OHDA must take into account differences in sensitivity of the different CA systems and stimulation of synthesis and utilization of serotonin (5-HT) following 6-OHDA administration. Thus, a chemical lesion of brain CA systems not only depresses the activity of these systems but may also induce compensatory mechanisms in the same systems (for instance postsynaptic supersensitivity, sprouting, etc.) and in other systems (disinhibition of 5-HT systems, modifications of endocrine systems).

Animal experiments involving chemical or surgical destruction of brain CA neurons can hardly be extrapolated to man. Some human pathological conditions, such as Steele–Richardson disease, involve extensive lesions of these systems. In the study of three cases of this disease, Leygonie et al.[10] described a marked decrease of PS, whereas its latency remained in the normal range. Rapid eye movements were present in slow sleep and slow waves appeared sometimes in PS. Abnormalities of PS have been described in Parkinson's disease[6,7].

Besides direct lesions of brain CA systems, there are basically five means to modify the activity of brain CA systems: inhibition of CA synthesis; modification of the availability of CA precursors; release of presynaptic CA stores by reserpine or other similar substances; modification of the physiological release of CA at the presynaptic level and modification of the sensitivity of postsynaptic receptors. Experiments performed with various drugs, aimed at testing one or more of these mechanisms, often led to contradictory results. In this discussion we shall focus primarily on pre- and postsynaptic components.

Inhibition of CA synthesis by αMPT is difficult to achieve because of the toxicity of this product. In man it can be given only at a relatively low dosage. In these conditions it has been reported to enhance PS[21]. In monkeys, Weitzman et al.[20] found a selective decrease of PS after αMPT, compensated by an enhancement of the other sleep stages and accompanied by a slight decrease of waking. Other experiments in cats, however, did not confirm these findings. In this animal, 75–100 mg of αMPT per kg induced a reduction of regional brain levels of NE in an apparently good correlation with an elevated amount of PS[8, 16].

In another experiment, αMPT was administered in successive injections of 75 mg per kg every 4 hours in order to avoid toxic reactions. The animals were continuously recorded and the general trend of PS was calculated[4]. In parallel experiments the fluorescence of brain CA was estimated. After a total dose of 75–150 mg of αMPT per kg, an enhancement of the production of PS was observed, while the decrease of brain CA fluorescence was only minimal; on the contrary, for higher doses of αMPT a substantial and progressive decrease of brain fluorescence occurred, in good correlation with a dose-dependent decrease of the production of PS. This finding can account for some previous contradictions by indicating that the effect of αMPT is dependent on the dose. Moreover, after low doses of the substance, the number of PS phases was enhanced, whereas their average length was unmodified or slightly decreased. A possible explanation could be that a slight inhibition of CA synthesis is sufficient to release a PS trigger mechanism normally held under tonic inhibition but insufficient to impair markedly the occurrence of PS. The net result is an enhancement of its production. Higher doses of αMPT lower the activity of brain CA synapses in such a way that the occurrence of this stage is impaired and its production decreases.

The administration of clonidine to rats leads to similar observations. We have given this drug intraperitoneally to chronically implanted rats at doses ranging from 2.5 to 20 μg per kg. This substance has been chosen because it is an alpha-adrenergic agonist, decreasing the stimulation-induced release of NE from central NE neurons. It reduces the turnover of NE in the brain and inhibits the spontaneous firing of NE cells in the locus caeruleus[18]. All tested doses induced an initial and dose-dependent suppression of PS. In later parts of sleep, PS reappeared at a higher rate of production than in control recordings: the slope of PS was significantly steeper than after saline. In addition, as was the case with αMPT, the number of PS episodes was increased. After the smallest dose of 2.5 μg per kg, the late enhancement of the production of PS was larger than the early suppression of this stage, resulting in an overall quantity of PS, counted in minutes, larger than in control recordings. The difference, however, was not significant due to inter-individual variability. This effect can be described as an intrasleep rebound.

In normal human subjects, clonidine at the dose of 0.4 μg per kg also induced in some subjects an enhancement of the production of PS. However, this effect was not present in all subjects studied in this way and, on the average, there was no difference with respect to placebo condition. Higher doses of clonidine resulted in a dose-dependent decrease of the production of PS, with some indications of an intrasleep rebound in the late part of sleep[1].

These results indicate that inhibition of CA synthesis and decrease of NE release by presynaptic stimulation of alpha-receptors influence PS in much the same way: the production of the stage is decreased, but the lowered activity of brain CA systems probably unmasks some trigger mechanism which becomes more active and can account for the more frequent appearance of PS. The identity of this probably non-CA primer mechanism cannot be determined at present. Two possible candidates can be mentioned: brain 5-HT systems could be involved according to Jouvet[3], and recent experiments in man indicate that an activation of cholinergic structures in the brain significantly shortens the latency of PS[15].

Neuroleptic drugs such as chlorpromazine are known as potent alpha-adrenergic blockers in the periphery as well as in the brain. When administered to rats at doses higher than 1 mg per kg, chlorpromazine decreases in a dose-dependent manner the production of PS, whereas at lower doses the production of PS is stimulated and results at the end of the recording period in a significantly larger quantity of PS[5]. In this case, however, the number of PS episodes is not increased as it is after clonidine. In other words, PS is not more frequently triggered, but its occurrence is maintained in each episode for a longer lapse of time. In addition, a pretreatment with a small dose of αMPT or with Fla 63, a dopamine-beta-hydroxylase inhibitor, completely reverses the PS enhancing effect of a small dose of chlorpromazine; in this condition, the production of this stage is markedly depressed[5]. Taken together, these results indicate that a small dose of chlorpromazine enhances the activity of CA synapses, probably by presynaptic blockade of alpha-receptors, leading to enhanced release of the transmitter. When the synthesis of the transmitter is slightly impaired, this blockade results in a rapid exhaustion of the presynaptic stores, and the activity of CA synapses is subsequently depressed.

In man, chlorpromazine has similar effects, stimulating the production of PS at low doses and depressing it at doses larger than 1 mg per kg. Moreover, the enhancement of PS under a low dose of chlorpromazine is completely suppressed by a small dose of clonidine, ineffective when given alone[1]. In this situation, clonidine does not exactly mimic the effects of αMPT; the combined treatment with chlorpromazine and clonidine does not depress markedly the production of PS as does the combination of αMPT and chlorpromazine in the rat, but only suppresses the chlorpromazine-induced enhancement. In fact, the general trend of PS under the combination of chlorpromazine and clonidine is superimposed on the trend of PS in baseline nights.

The experimental data obtained in the rat and in man summarized here point to a positive correlation between the activity of CA systems in the brain and the production of PS. The converse explanation appears very unlikely as it does not account for the biphasic effect of chlorpromazine and of αMPT. Also the idea of a negative correlation would not provide an explanation for the effects of combinations of αMPT and chlorpromazine, and of clonidine and chlorpromazine respectively.

If intact synaptic transmission in CA neurons is necessary for the realization of waking and PS, we are left with three major questions, pointing to further lines of research. The first problem is related to the common involvement of CA neurons in waking and in PS; this leads us to suspect the existence of other functional or biochemical differences between these two states. These other factors are at present unknown. It should be kept in mind that brain CA structures are not isolated but extensively interconnected with other systems. The study of these connections is not in an advanced stage but would contribute to a better understanding of these mechanisms.

The second problem is related to the fact that there is no simple relationship between the level of endogenous monoamines and the states of vigilance. As we have seen, chemical lesions with 6-OHDA or depletion of endogenous stores with reserpine are followed by a progressive recovery of PS while brain monoamines are still largely depleted. A possible explanation of this discrepancy could lie in the adaptative and replacement capabilities of the central nervous system.

The final problem we would like to mention here is the effect of antidepressant drugs on sleep. Most of them depress markedly PS, an effect which is, at the first glance, difficult to reconcile with one of their best-established biochemical properties, namely the inhibition of the reuptake of monoamines. A first indication is given by the recent description of other antidepressant substances which do not depress PS. It appears likely that there is no direct and simple relationship between the depression of PS and the antidepressant properties. There is, at present, no satisfactory explanation of the effects of these substances on sleep, but we feel that such an investigation would be an important step in the understanding of their effects on mood regulation.

## References

1. Gaillard, J.-M. and Kafi, S. (1979). Involvement of pre- and postsynaptic receptors in catecholaminergic control of paradoxical sleep in man. *Eur. J. Clin. Pharmacol.* (In press)
2. Hartmann, E., Chung, R., Draskoczy, P. R. and Schildkraut, J. J. (1971). Effects of 6-hydroxy-dopamine on sleep in the rat. *Nature (London)*, **233**, 425
3. Jouvet, M. (1969). Biogenic amines and the states of sleep. *Science*, **163**, 32
4. Kafi, S., Bouras, C., Constantinidis, J. and Gaillard, J.-M. (1977). Paradoxical sleep and brain catecholamines in the rat after single and repeated administration of alpha-methyl-paratyrosine. *Brain Res.*, **135**, 123–134
5. Kafi, S. and Gaillard, J.-M. (1978). Biphasic effect of chlorpromazine on rat paradoxical sleep: a study of dose-related mechanisms. *Eur. J. Clin. Pharmacol.*, **49**, 251
6. Kales, A., Ansel, R. D., Markham, C. H., Scharf, M. B. and Tjiauw-Ling Tan (1971). Sleep in patients with Parkinson's disease and normal subjects prior to and following levodopa administration. *Clin. Pharmacol. Ther.*, **12**, 397

7. Kendel, K., Beck, U., Wita, C., Hohneck, E. and Zimmermann, H. (1972). Der Einfluss von L-dopa auf den Nachtschlaf bei Patienten mit Parkinson-Syndrom. *Arch. Psychiatr. Nervenkr.*, **216**, 82
8. King, C. D. and Jewett, R. (1971). The effects of alpha-methyl-tyrosine on sleep and brain norepinephrine in cats. *J. Pharmacol. Exp. Ther.*, **177**, 188
9. Laguzzi, R., Petitjean, F., Pujol, J. F. and Jouvet, M. (1972). Effets de l'injection intraventriculaire de 6-hydroxy-dopamine. II. Sur le cycle veille-sommeil du chat. *Brain Res.*, **48**, 295–310
10. Leygonie, F., Thomas, J., Degos, J. D., Bouchareine, A. and Barbizet, J. (1976). Troubles du sommeil dans la maladie de Steele–Richardson. Etude polygraphique de 3 cas. *Rev. Neurol.*, **132**, 125
11. Lidbrink, P. (1974). The effect of lesions of ascending noradrenaline pathways on sleep and waking in the rat. *Brain Res.*, **74**, 19–40
12. Matsumaya, S., Coindet, J. and Mouret, J. (1973). 6-Hydroxydopamine intra-cisternale et sommeil chez le rat. *Brain Res.*, **57**, 85–95
13. Matsumoto, J. and Jouvet, M. (1964). Effets de réserpine, DOPA et 5-HT sur les deux états de sommeil. *C. R. Soc. Biol.*, **158**, 2137
14. Pujol, J. F., Kan, J. P., Buda, M., Petitjean, F., Mouret, J. and Jouvet, M. (1975). Is 6-hydroxy-dopamine (6-OHDA) a specific tool for the study of functional roles of catecholaminergic (CA) neurons in the sleep–waking cycle? In G. Johnson, T. Malmfors and C. Sachs (eds.), *Chemical Tools in Catecholamine Research*, Vol. I, pp. 259–266. (Amsterdam: North-Holland Publishing Company)
15. Sitaram, N., Moore, A. N. and Gillin, J. C. (1978). Induction and resetting of REM sleep rhythm in normal man by arecholine: blockade by scopolamine. *Sleep*, **1**, 83
16. Stern, W. C. and Morgane, P. J. (1973). Effects of alpha-methyl-tyrosine on REM sleep and brain amine levels in the cat. *Biol. Psychiatry*, **6**, 301
17. Stern, W. C. and Morgane, P. J. (1973). Effects of reserpine on sleep and brain biogenic amine levels in the cat. *Psychopharmacologia (Berlin)*, **28**, 275–286
18. Svensson, T. H., Bunney, B. S. and Aghajanian, G. K. (1975). Inhibition of both noradrenergic and serotonergic neurons in brain by the a-adrenergic agonist clonidine. *Brain Res.*, **92**, 291–306
19. Torda, C. (1968). Effect of changes of brain norepinephrine content on sleep cycle in rat. *Brain Res.*, **10**, 200–207
20. Weitzman, E. D., McGregor, P., Moore, C. and Jacoby, J. (1969). The effect of alpha-methyl-para-tyrosine on sleep patterns of the monkey. *Life Sci.*, **8**, 751
21. Wyatt, R. J. (1972). The serotonin-catecholamine-dream bicycle: a clinical study. *Biol. Psychiatry*, **5**, 33

# 5

# Melatonin and Sleep in Man: a Preliminary Report*

## T. Hansen, A. J. Birkeland and O. Lingjærde

In northern Norway there are very marked seasonal variations in light intensity, with the sun above the horizon day and night for several weeks in the summer, and below the horizon for a similar period in the winter – the 'season of obscuration'. Therefore, there are excellent opportunities for studying circadian and seasonal variations in biological functions in northern Norway.

Many people suffer from insomnia in the season of obscuration, some also in the midnight-sun period.

We have now started to investigate the circadian profile of serum melatonin in relation to sleep in six male volunteers at four different times of the year (January, April, July and October). In each period the subjects sleep for two consecutive nights in our sleep laboratory. After the first night, blood samples are drawn from an indwelling venous catheter with one-hour intervals from 9.0 a.m. to 11.0 p.m., and with 20-minute intervals from 11.0 p.m. to 8.0 a.m. the next morning. Sleep is recorded polygraphically, and the sleep stages assessed according to the criteria of Rechtschaffen and Kales[1]. Serum melatonin is determined by the RIA method of Rollag and Niswander[2].

---

*As this is only a preliminary report of current studies, publication of the full results will appear elsewhere.

In this preliminary communication, we present the melatonin and sleep curves from four persons from January 1978.

The 24-hour melatonin curves show marked individual variations, and a rather irregular pattern with more or less marked peaks mainly in the afternoon and during the night. Overall, the melatonin level is higher during the night than during the day. When seen in relation to the sleep pattern, the most conspicuous feature is a fall in serum melatonin upon onset of sleep, and raising levels in connection with waking periods during the night and at the time of waking up in the morning. In some of the curves, low levels seem to be connected with REM sleep, but on the whole, so far we have found no *certain* connection between serum melatonin and the different sleep stages. At this stage, neither can we say to what extent the serum melatonin pattern is influenced by the season of obscuration.

Our results seem to indicate that melatonin secretion is not stimulated by sleep as such – rather on the contrary. Melatonin may be more important in the process of *inducing* sleep than in *maintaining* sleep.

## References

1. Rechtschaffen, A. and Kales, A. (1968). *A Manual for Standardized Terminology, Techniques and Scoring System for Sleep Stages of Human Subjects*. Public Health Publications No. 204. (Washington, D.C.: U.S. Government Printing Office)
2. Rollag, M. D. and Niswander, G. D. (1976). Radioimmunoassay of serum concentrations of melatonin in sheep exposed to different lighting regimes. *Endocrinology*, **98**, 482

# Discussion I

## Moderator: Professor R. G. Priest

**Priest:** Dr Möhler, you've provided very convincing evidence that there is a specific benzodiazepine receptor. Where does research go next, do you think?

**Möhler:** Research in several laboratories, including our own, is attempting to find the endogenous ligand, and that I think will reveal new aspects in all disorders which are ameliorated by benzodiazepine treatment.

**Marks (Cambridge):** Dr Möhler, you have shown that there is a very high affinity for benzodiazepines within the cerebral cortex. Have you studied various areas of the cortex and, if so, are there any differences between the different areas?

**Möhler:** We have studied three areas of the cortex and they all had the same affinity. Others have studied the cortex in even more detail and they didn't find any differences in affinity of the receptor, nor were there great differences in the density of the receptor in different cortical areas.

**Saletu (Vienna):** How about the affinity to other tissues, for instance, cardiac tissue?

**Möhler:** This hasn't been studied *in vivo*, but *in vitro* GABA and the benzodiazepines occupy different receptors and they don't interact with each other at their own binding site. Recent *in vitro* experiments by Dr Costa at the National Institute of Mental Health show that there may be an interaction between these two receptor systems, and that if a ligand is bound to the benzodiazepine receptor it may alter the characteristics of a neighbouring GABA receptor. This is one theory which is at present under investigation in the laboratories. Our idea, however, would be somewhat in contradiction to this because we didn't find a single case in our studies in neurodegenerative diseases in which a benzodiazepine receptor was associated with a GABA receptor. We don't feel that all benzodiazepine receptors are sitting so to speak next to a GABA receptor. There are at least some benzodiazepine receptors which are there by themselves and have a function as yet unknown. Perhaps they are just waiting for the endogenous ligand. These are the two theories at present.

45

**Lader:** I was interested in your experiments with Huntington's chorea and I wondered if you had thought of doing the same sort of experiment in post-mortem tissue from patients with dementia where there is very good evidence of cortical drop out of cells that would fit in with the very high receptor populations in the cortex.

**Möhler:** Yes, we would very much like to do this. We have just done a study in Parkinson's disease and would like to go on to look at tissue from other neurodegenerative diseases. The problem is to have well-defined dementia patients and tissue.

**Erdmann (Groningen [formerly Birmingham, Alabama]):** That was very interesting, Dr Ingvar, what you just explained about cerebral blood flow and sleep. We have done some experiments with oxygen diffusion coefficients in sleeping monkeys and we observed a vast decrease of the oxygen diffusion coefficient in the tissue during sleep which cannot easily be explained. Probably an increase of cerebral blood flow is just a compensation for the decrease of oxygen diffusion coefficient in the tissue, so it might be an explanation for our results – which we never published because we didn't believe in it.

**Ingvar:** It is a pity that you didn't publish your experiments, but you are right about the complexity of the relationships between the oxygen tension, oxygen availability, functional activity and blood flow. I think you are right in your interpretation that the change in flow which you have when you go to sleep, which in part is due to the respiratory increase in carbon dioxide tension, probably explains your findings, but I think you should publish them.

**Chruschiel (Warsaw):** I'm a little bit worried about the use of clonidine in such experiments because clonidine has been recently shown to be involved in some way with endorphin receptors too. It is a multispectrum drug and it influences not only the alpha activity but also probably, as a kind of semi-ligand, the endorphin receptor activity. Therefore your very interesting results might be explained partially, to my mind, by the possible effect of clonidine upon other receptors too, not only the alpha-adrenoceptor itself.

**Ingvar:** Some years ago we showed that the melanocyte-stimulating hormone (MSH) has a synchronizing effect in man on the EEG. Professor Lingjærde, have you any comment on how that result could be related to your report? Secondly, how do your very interesting findings relate to other hormonal studies at night? Is there any relation?

**Lingjærde:** I have no comment on the melanocyte-stimulating hormone. As to the other hormones, we plan to measure some of them in the same blood sample – for instance, prolactin. But so far we haven't done that. A study was done in Tromsö several years ago when cortisol and growth hormone were measured in plasma, in collaboration with Dr Weitzman in New York. That was mainly to see if there were some seasonal variations, and they also measured it at four different times of the year. They didn't find any marked seasonal variations – only the usual pattern for these hormones.

**Oswald:** Professor Lingjærde, in your preliminary remarks I think you said that the synthesis of melatonin was increased during the dark. Can I ask whether this relates to the darkness and light or does it relate to the activity–rest pattern of the animal, because some animals are nocturnal, some diurnal?

**Lingjærde:** I think that it has been shown in rats that darkness as such seems to be the stimulating factor, and you can interrupt the increase during the night by having constant light throughout. If you have constant darkness throughout day and night, you still see the circadian variation in serum melatonin in rats, but it may be somewhat different in humans. I think some studies indicate that activity *and* light *and* sleep as such may have an influence on the synthesis rate, but so far too little is known about that, especially in man.

**Lader:** Professor Ingvar, in your technique, are you measuring the blood flow just on the surface or is it deep blood flow? Secondly, does the technique you've been using differ from the positron detection technique, which I suspect does show the blood flow rather more deeply?

**Ingvar:** We look right through the hemisphere which is being labelled with the isotope. But due to absorption of this weak gamma-radiation from xenon-133, the charts I showed are mainly charts of the surface blood flow so that each detector looks right through the hemisphere and sees a truncated cone where the outer part is seen about three times as well as the mesial part.

The positron camera has the unique feature that it looks right through the head and there is no superposition, it's a real three-dimensional picture of the flow. The 'pig' cells, which is the technical term for size of the units which can be differentiated, have a size of some one cubic centimetre and they are not superimposed. It's like in the CAAT scan, you really look at the flow and the metabolism in that very small volume of the brain in depth and this affords simply fantastic possibilities to look at the exact cerebral metabolic correlates to individual mental acts, which is quite a new thing. It is my conviction that this technique will dominate future research, not only in sleep but very much so in psychophysiology and psychopathology, too. Already there is some evidence that some mental disorders are accompanied by sleep disturbances. Chronic schizophrenia, paranoid states, depression and to some extent manic states have abnormal functional distributions which have not been known previously, and these can now be studied with the positron technique, which is, however, only available at very large centres in the United States and one in

Stockholm. The prerequisite for using these methods is that you have to have a cyclotron in the basement, and a cyclotron is a rather costly thing – a matter of some 3 million dollars.

**Priest:** Am I right in thinking that the instrument works on a similar principle to the Emiscan except that it is an emission rather than a transmission?

**Ingvar:** Exactly. The image reconstruction as grids – the density diagram in a slice of the brain – is theoretically extremely closely related to the same grids used in transmission tomography. So we now have two instruments, one for the remarkable morphological studies of the brain *in situ*, a dramatic technique, and in addition to this the positron to study the metabolism. I should add perhaps that you can use oxygen-15 to get oxygen consumption, you can use fluorodioxoglucose to get glucose consumption, you can use ammonia, you can use a number of substances, and you can also use labelled drugs and see where they go, which indeed gives a fascinating prospect.

**Oswald:** Professor Ingvar, in your talk you mentioned the paper by Mangold *et al.* and pointed out their awkward finding that there was no mean fall in cerebral blood flow during human sleep. I was not at the end clear from your own talk whether you were concluding, as I hope you will because it would suit me, that cerebral blood flow in man is lower during sleep. I would say that the majority of studies suggest that cerebral blood flow during human REM sleep is higher than during wakeful rest, but my question is, have you satisfied yourself that it is lower than it is in wakefulness? I hope you have.

**Ingvar:** Well, Professor Oswald, I was trying to obscure the issue but I didn't succeed, apparently. I think there is some evidence that the net total mean cerebral blood flow is slightly lower in slow-wave sleep than in wakefulness. This is one thing that I think there is evidence for. However, we have occasional accidental observations which indicate that very marked redistributions take place. In the future, we shall have more information on these regional distributions, which I think are the critical points. It appears as if the blood volume, the only relative measure we have, is higher in many cortical areas in REM sleep than it is in wakefulness and also in slow-wave sleep. So this would be in line with the evidence from animal experiments, that there is a moderate increase in the total net cerebral metabolism in the REM phase. However, again one may have increases which are very much more marked in some regions than in others. But we do not know the regional distribution as yet.

**Oswald:** Yes, but the critical question is, is there an increase in metabolism to go, as you seem to assume, with that increased flow in REM sleep? Reivich showed that blood flow in the red muscles in REM sleep is cut to about one-third that of slow-wave sleep, whereas this does not happen in anaesthesia. That is to say it is not a result of lowered work in the muscles, it must be some positive neurogenic phenomenon unrelated to metabolism. Now might it not also be true in the brain that this raised blood flow during REM sleep has nothing to do with metabolism?

**Ingvar:** That is a very good point indeed, that remarkable increase in flow which has been so well demonstrated in animals by the Reivich group. I showed one slide which indicated that the mean increase of flow is about 80%. This has always been astonishing to many people and indeed in the explanation of this phenomenon people have mentioned the vasomotor factor as one possibility, which would certainly take this increase out of the usual coupling story between neuronal function and flow. I think that this is certainly a possibility – the nerve fibres are there around the brain vessels; but the solid evidence that these play any role under normal conditions is very scanty, so I'm no great believer in the vasomotor alternative. But we shall simply have to do more work to find out.

**Oswald:** But we might remember that in REM sleep the penis becomes erect without, one supposes, any vital metabolic basis.

**Ingvar:** How do you know? *(Laughter.)*

**Oswald:** O come on, you know that too, and everybody. Well, perhaps it does. That's all, it's just an illustration.

**Scherschlicht (Basle):** I wanted to ask Professor Ingvar about the velocity of these changes in blood flow when you say to a patient (or volunteer) 'Imagine, you are moving your hand', for instance?

**Ingvar:** The time resolution of these measurements is very low. It's a question of 2 minutes at least, which means that we cannot at all follow rapid changes in activity, such as the rapid landscape changes which accompany, for example, ordinary speech, ordinary thinking if you like. We are very far away from looking at what Sherrington once called the 'enchanted loom' when he was writing about the signals inside the neuropile of the cortex which run in and out like small lights. This enchanted loom we cannot see by far because these methods are so slow at present. We can get down somewhat by using the technique worked out by Cooper and Gray in Bristol by which they could follow with implanted electrodes some of the metabolic and flow phenomena in the cerebral cortex during mental work. But the rapid EEG phenomena of work potentials and so on are simply on quite a different time scale of milliseconds, while we are working on minutes.

**Saletu:** Dr Gaillard, I wonder whether you could comment on brain catecholaminergic activity and other classes of drugs, for instance the antidepressants, as there are now selective catecholamine-reuptake inhibiting drugs available – or, let's say, antidepressants which inhibit more the catecholamine reuptake, such as mianserin. Now you have shown that with the neuroleptic chlorpromazine in low dosages you get an enhancement, and in high dosages a suppression, of REM sleep. What is the effect of such specific reuptake inhibitors?

**Gaillard:** We have not yet done similar experiments with antidepressant drugs, but you are right, this is an important problem. However, we should not forget that while some antidepressant drugs are potent inhibitors of reuptake, they also do many other things. In particular, they have anticholinergic activity and this could explain some of the phenomena. On the other hand, more recently other antidepressant drugs have been described which do not depress, or at least do not seem to depress, the production of paradoxical sleep. I'm not completely sure, but I think mianserin does not depress REM sleep, or at least not markedly.

**Monti (Italy):** Professor Ingvar, it is known that several drugs like the benzodiazepines induce sleep, but they change the normal pattern, that is, slow-wave sleep decreases. Are the blood flow patterns different if you compare physiological sleep with drug-induced sleep, or are they the same?

**Ingvar:** I am sorry, but we have no information on drugs and blood flow. We have much information on EEG and drugs, but I don't think we should mix this into it. This much I can say, there is every reason to believe that the landscapes I have shown have a correspondence in the EEG picture. So we are starting to ask the EEG new types of questions which are mainly concerned with the regional distribution of EEG pattern, and possibly we can in this way also get some answers to whether drugs would induce a different pattern or distribution of activity in the brain than without drugs.

**Marks:** Dr Gaillard, you showed some cumulative REM phases in humans. Are those on single night observations? If so, how much variation is there between individual nights, without any interference?

**Gaillard:** The curves I have shown, representing the general trend of REM sleep, are always averaged on several subjects, 6–12 subjects. And I can say, for instance, that the variability is not very great in control nights or after drug studies, for instance, with a small dose of chlorpromazine. This dose has been studied in 12 subjects, and all 12 subjects showed some enhancement of the production of paradoxical sleep. With a lower dose of 0.25 mg per kg there was a greater variability. We have studied this dosage in six subjects, and three of them showed in the initial part of the night an enhancement of paradoxical sleep, whereas the three other subjects did not show this effect. This can probably be related to the variability in the pharmacokinetics of this drug in different individuals. For clonidine also, the effects were rather similar in all subjects. With the highest dose of clonidine we have used, all subjects showed a decrease of the production of paradoxical sleep.

**Lehmann (Zurich):** Dr Ingvar, in your demented subject who went to sleep, you showed a regional increase in the lower temporal lobe, which was not present in the wake stage and seemed to be the difference between wake and sleep. What kind of activity would this correspond to in a normal person?

**Ingvar:** In fact, the information we have on the temporal lobe is limited. We have studied some 90 dementia patients in detail, with flow studies, psychometric techniques and so on, and have some evidence to confirm the old idea that the temporal lobes have to do with memories. These memory defects are correlated with a very low flow in the temporal lobe. Now, as for this accidental observation by us and by a few others, that temporal lobe activity increases during slow-wave sleep, we have no real idea what it means: we can only guess. Probably the normal activity pattern of wakefulness, with a high flow in the frontal areas and low on the afferent side, if you like, around the auditory, the visual and somato-sensory regions, signals a type of activity of the brain which is probably rather critical for wakefulness.

When we are awake there is an inhibition on the afferent side, so that we are not bothered by the Niagara of afferent impulses which constantly impinge on us; at the same time we are continually producing or simulating some anticipatory behaviour. As we see it, this is probably the essence of wakefulness or of consciousness. Now these few observations would perhaps signal that this control on the input side is removed in sleep. This is a very loose hypothesis and you can take it for what it is, but at least the reversal of the wakefulness pattern might signal such a very fundamental change in brain activity in which the control on the input goes away and perhaps the production of behavioural motor patterns also stops in sleep.

**Erdmann:** I just want to come back to these cerebral blood flow changes. I think that we have not only to see the cerebral blood flow in isolation, we have to look into the intercapillary tissue. Oxygen is coming to the cells not only via the blood flow but also by diffusion from the capillaries to the cells. Now if the resistance of the intercapillary tissues to diffusion increases during sleep, less oxygen arrives in the cells far away from the capillary, which would mean a feedback to increase the blood flow. We try to explain this phenomenon by the hyaluronidase mucopolysaccharate mechanism which changes diffusion resistance, and we are working on this hypothesis with microelectrode studies.

**Ingvar:** To my mind there is no reason to believe that we have any substantial changes in the oxygen diffusion constants of the tissue when the functional activity of the tissue changes. Oxygen diffusion is, after all, determined by the physicochemical characteristics of the tissue as such. This freely diffusing gas goes right through nervous tissue whether this is active or not, so I am not ready to believe that alterations in the diffusibility of oxygen would have any importance for the induction of sleep.

**Erdmann:** I think quite differently because some investigations, which we published last year, showed that the oxygen diffusion coefficient changes markedly according to the condition of the tissue. For instance, in ischaemia the oxygen diffusion coefficient decreases from 1.8 to 1.4 × $10^{-5}$.

**Ingvar:** Our studies in monkeys during sleep, which were just control studies of the diffusion coefficient over a longer period, showed that it also goes down during sleep. It seems, from the literature and also from what Dr Putkonen said, that whatever drugs you give, in high enough doses, suppress REM sleep at least for some time, then you may have a rebound later on, or at least, a return to normal. I wonder, is it really useful to try to deduce too much information from REM sleep, paradoxical sleep reduction under drugs? On the other hand, an increase could probably tell you something.

**Putkonen:** Well, I think it is useful in spite of what Hartmann said in Montpellier because he didn't like the data that I and Gaillard were presenting. You see, not every drug decreases REM sleep. You can give tremendous amounts of phentolamine for instance, and this drug increases REM sleep, and with chloropromazine you can increase it too, and I think you have to look at the whole set-up.

**Björqvist (Finland):** I have been looking at alcohol myself, and the most impressive findings are the changes in slow-wave sleep that come immediately after taking this particular drug. I wonder if other drugs cause changes in K-complexes and spindles and so on, during slow-wave sleep, that are obscured by looking only at sleep stages.

**Gaillard:** Benzodiazepines decrease the production of slow-wave sleep, decrease K-complexes and enhance spindles, roughly speaking. Another peculiarity of these drugs is that the time-course of the effect on paradoxical sleep and on slow-wave sleep is not the same. Paradoxical sleep is depressed by benzodiazepines, depending on the dose, and recovers quickly after interruption of the treatment, whereas slow-wave sleep is influenced for a much longer lapse of time. There must be some difference of action in the brain system responsible for these stages, but we do not understand these differences yet. With clonidine we have also rather a strong effect on slow-wave sleep and one important point would be to study the relationships between the effect on slow-wave sleep and on REM sleep, but it's a difficult point.

**Scherschlicht:** I should like to make a short comment on REM suppression with special respect to antidepressants. I have no experience with man, but in cats we are able to suppress REM sleep with very low doses of antidepressants. You can indeed suppress REM sleep with every substance, depending on the dose, but this depression of REM sleep in the cat with antidepressants is produced by very low doses which do not affect any other sleep parameters and we use this special feature for the screening of antidepressants.

**Monnier (Basle):** Could you comment on the critical level of reduction of blood flow, blood pressure, or respiration for metabolism? Humanity is fed today with tranquillizers, hypnotics and beta-blockers, so that it could easily happen that a critical dose would be reached. Do we have data on animals which allow you to say: here it becomes critical for neuronal metabolism?

**Ingvar:** In answer to Professor Monnier's question, I can say that first, from studies mainly carried out in Professor Bo Siesjö's laboratory, we can say that the effect of clinical dosages of sedatives and neuroleptics on the total net metabolism is very small, indeed hardly measurable. But it is probably the effects upon critical parts of the brain which lie behind the very well-established clinical effects. Of course, the situation becomes very different in intoxications. In barbiturate intoxications you do get down to very low levels. I mentioned, in my paper (Chapter 2), that the normal metabolism lies around 3.3. With high doses of barbiturates, you can easily come down to 0.5 or even lower, and then you reach these thresholds that you were asking about. I think it's fair to say that we have two thresholds. One is the electrical threshold where the blood flows through the cortex; this is in the vicinity of about 16–20 instead of the normal 100. Below 16, you come to the next threshold and that is the ischaemic threshold. These are figures established in studies by Lindsay Simon at the National Hospital in London, amongst other people. So these thresholds are two, shall we say, very important values to know about, because that is where the critical threshold lies. We can find them, we can to some extent identify them, in ischaemic brain disorders, in stroke patients. The centre of the stroke where the neurons are dead has a flow which is lower. The surrounding tissue may have a flow between 16 and 25, this is that zone which we try to affect by drugs with shunt operations. But you are right, there are definite thresholds for the flow, and below those thresholds we endanger the life of the neuron.

**Priest:** To what extent is the secretion of melatonin continuous and to what extent pulsating?

**Lingjærde:** I think the synthesis of melatonin in the pineal is continuous during the dark period, but the secretion from the pineal to the blood seems to go in bursts. This is just what you find, for instance, for hydro-cortisone, which also seems to come in peaks, and it may perhaps be true for many hormones. I think this peak pattern has not been shown before to be so very marked for melatonin.

**Hartelius (Kristianstad):** Professor Ingvar, in REM sleep there are dreams, and I suppose there must be differences in quality of dreams. I myself experience some comfortable dreams and also sometimes dreams that wake me up, and I suppose there must be corresponding differences in the metabolism. Also it is rather easy to get the patient to assess the quality of the night. Have you made any annotations to that kind? Secondly, also from a clinician's point of view: we know that we have oneirophrenic hallucinations with a rather low level of activity, and we have hallucinations during excitation. There is a big difference between a delirium tremens patient with anxiety up to agony and those just dreaming or something like that, with rather little clinical activity.

**Ingvar:** We have no information on whether there are different blood flow landscapes for different dreams. We simply have not had an opportunity to study this. We'll have to wait for new and less traumatic techniques to go into this extremely interesting story. As far as hallucinations are concerned, we and others have evidence that these techniques are sufficiently sensitive to look at the pathophysiological substrate for hallucinations. My former co-worker who is now independent, John Risberg, recently had a case of alcoholic hallucinosis. It is an interesting story about a young Swedish man who had been drinking alcohol with his brother for 3 weeks. He stopped drinking because the brother went home, but he then observed that he was constantly hearing the Swedish national anthem being sung by a choir, and this was very disturbing to him. *(Laughter.)* So he went to the psychiatrist at Lund and complained about that Swedish national anthem which he constantly heard following his drinking bout. And they made a flow study, which showed a very marked regional flow in the auditory and para-auditory regions.

So this fellow was given haloperidol or something, and next day he said: 'Oh, Doctor, I am so much better, now I hear the music but not the text'. *(More laughter.)* So they made another study and the flow certainly had gone down but you could still see a slight hyperaemia in this temporal region. Well you know the end of the story was that the next day he said: 'Doctor, now it has stopped, all of it', and there was no hyperaemia left.

It certainly is a very good story but the funny thing about it is that it is true. We have other similar patients and believe we can really measure the intensity of the hallucinatory activity by looking at the blood flow. We also have some evidence from studies on schizophrenics that the disturbed cognition and the hallucinatory activity which these patients experience correspond to increased activity in the posterior parts.

**Oswald:** Dr Möhler, some of us have to teach young psychiatrists who are training and want to pass examinations: and among the things they think they will be asked is 'What are the differences between benzodiazepines and barbiturates in their actions on the brain?' The standard reply that they have come to believe they should offer is 'Benzodiazepines act upon the limbic system'; but you showed us that benzodiazepines have more receptors in the cortex. What should students now reply?

**Möhler:** I guess it is the old story, the questions remain the same, the answers differ. The old answer about the limbic system stems from electrophysiological experiments where the first potential changes after administration of a benzodiazepine could be picked up in the limbic system. But in view of the benzodiazepine receptor distribution in the brain one would assume that other areas, such as the cortex, are equally if not more important. What the barbiturates are doing is probably very similar to what the benzodiazepines are doing as far as GABA-ergic mechanisms are concerned, but there is one clear-cut difference, they don't act via the benzodiazepine receptor. It has been proposed recently that barbiturates may actually block the chloride channel activated by GABA neurons. In other words, the opening time of the chloride channel in response to stimulation of the GABA receptor may be regulated by the barbiturates. The benzodiazepines on the other hand act via the benzodiazepine receptor, which may be located either presynaptically on a GABA neuron or postsynaptically next to a GABA receptor. According to these two theories, this is the most clear-cut difference between the barbiturates and the benzodiazepines on a molecular level, but they both basically affect GABA-ergic systems.

**Ingvar:** You answered this question in a very satisfactory way, but I should like to ask you whether you are not paying too much attention to just the neurons. There is a report in a recent issue of *Science* about the action of the benzodiazepines upon the glial cells.

**Möhler:** There are conflicting reports. Some people say glial cells have benzodiazepine receptors; some people, like our colleagues Braestrup and Squires in Denmark, didn't find any, so it may depend on the type of glial cell or the type of tissue culture you are looking at. It is also an open question whether this is of functional significance in the brain.

**Oswald:** I should like to challenge what Professor Ingvar said, that we don't really have evidence that the benzodiazepines or other drugs affect brain metabolism. There are surely indications that I think are important for a meeting of this kind. I ask the question: What is sleep for? In all our cells there is continuous degradation and also there is synthesis, and the balance shifts during rest and sleep in at least most of the body cells, and that includes the brain. There are three papers showing that protein synthesis proceeds faster during the time of rest and sleep in the brain, as well as in a great many other tissues. And one must ask what do hypnotic drugs, for example, do to this balance between synthesis and degradation. The evidence overall suggests that in clinical doses they assist the process of synthesis. You yourself showed us a picture indicating that cerebral blood flow is higher in anxiety. We know that in anxiety more epinephrine is secreted. This is a catabolic hormone. We know that in people who

55

sleep badly and are anxious, body temperatures are higher during the night. That suggests a higher rate of degradation. We know that they excrete more corticosteroid products in their urine. These are catabolic hormones. Now, my colleagues and I some years ago showed that a modern minor tranquillizer reduced the levels of corticosteroids during sleep, in blood sampled only during sleep. If we take account of these apparent hormonal effects, which are largely inferences I agree, we would think that modern minor tranquillizers and hypnotics would diminish the pressure towards catabolic activity during sleep. In addition, we know that they do increase the duration of sleep and diminish the periods of wakefulness, so that globally they might indeed exert an effect on overall metabolism.

**Ingvar:** Thank you, I think this was a most important comment. My sweeping statement that the drugs in ordinary clinical doses do not seem to affect the total net cerebral oxygen uptake appreciably should indeed be modified. I listened with great interest to your comment. It certainly contained several things which we could discuss. But we do know very little about this. However, we must remember that we most certainly have regional effects, both on the flow and on the metabolism, and that part of the beneficial and part of the not-so-beneficial effects of the various drugs should be referred to their effects upon regional neuronal systems. Protein synthesis certainly changes, hormonal patterns change, and you can affect this by drugs and secondarily by improving sleep, but the total net effects are still quite small. I was wondering if we could have a comment from Dr Bengt Nillsson from Dr Bo Siesjö's laboratory, where they have done very detailed studies of the various hypnotics, sedatives, tranquillizers, etc., in animals.

**Nillsson:** From our various experiments performed on rats, I can mainly confirm what Dr Ingvar said, that in the basic state of the animal chlorpromazine, benzodiazepines and so on don't affect the general metabolism very much. On the other hand, if you have the animals in a stress situation such as can be induced by what we call immobilization stress, or if you inject epinephrine, then you increase the metabolic rate in the brain substantially, and this can be abolished by, for instance, a benzodiazepine. This seems to be a more selective effect than, for instance, the effect you get with a barbiturate, which lowers metabolic rate generally.

**Ingvar:** May I add to Professor Oswald again that you are certainly correct that many patients with insomnia are also anxious, and this would, in our context, imply some slight but abnormal increase in the cerebral metabolic rate. The use of anxiolytic and sedative drugs in such patients may in fact reduce cerebral metabolism from a supranormal to a normal level.

**Oswald:** May I just remind people that the whole body is affected by sleep. Whole-body oxygen consumption falls very steeply during sleep and is lowest of all in slow-wave sleep. We mustn't just think of sleep affecting the brain alone, and our anxious patients, who may just feel bad all over, may be telling us something about their whole bodies when they are sleeping better.

**Ingvar:** Yes, and when they get less tense the activity in the body goes down, so does the sensory feedback from the tense muscles, and this may indeed also have regional effects upon the cortical fields.

**Billiard (Montpellier):** Dr Lingjærde, I was very much interested by the fact that 25% of the people in Tromsö experience sleeping difficulties in the dark period. Indeed I have met in the south of France a lot of people, in Marseilles by the way, who share a common experience that they have a different sleep schedule depending on whether they go northward or southward during the summer season, with more sleep in northern countries when the night is short and less sleep in southern countries when night is much longer than in northern countries. So I would like to know whether you think it would be of interest in addition to have subjects monitored in the darkness period and in the midnight sun period in your country, as well as in a southern country at a few days' interval, and to have periodic blood samples in both places.

Secondly, do you record the subjects in a dark room or in one lit by the outside light?

**Lingjærde:** To answer the second question first. Our sleeping laboratory is rather dark; there is only very dim light and there are no windows, so no external light. Of course, it would be of interest to study the same persons in different latitudes. There was an investigation in Norway several years ago when school children in northern Norway were compared with school children in southern Norway, where there is no period of obscuration. There was a very marked difference between the frequency of sleep disturbances, some 30% in northern Norway and only 5 or 6% in southern Norway, where people usually do not experience the same difficulties with sleep at particular seasons. There are other investigations which also indicate that some few people in northern Norway have sleep difficulties, especially in the fall and in the spring, but not in winter and summer. There are more people who have sleep difficulties in the winter and some of them also have difficulties in the middle of the summer. It is often said that people up north do not sleep during the midnight sun period; they can't afford to because there is so little light in the winter, so they must use what they have in the summer. I think actually people on an average sleep about the same number of hours in the summer as in the winter. The difficulty is mainly going to sleep in the evening. They just lie there and cannot sleep for several hours and this is quite monosymptomatic; it is not accompanied by depression or anxiety or bodily symptoms, just insomnia and nothing else. You can see it in small children and elderly people, perhaps more commonly in females than in males, and we found it was a little more common in young adults than in elderly people. Our experiments with melatonin are only one step to try to elucidate what is really behind this. Of course it would be interesting to do what you suggested in different latitudes.

**Laihinen (Finland):** Dr Möhler, the possibility was mentioned that there are endogenous benzodiazepine-like substances in the brain. Do you have any idea, or perhaps a speculation, as to what kind of substance this would be?

**Möhler:** We have several ideas ourselves and there is one suggestion, which is probably wrong, that hypoxanthine could be one of those compounds. But the problem is that all compounds isolated so far from brain tissue which could serve as endogenous ligands of the benzodiazepine receptor have a very low affinity to this receptor. Now this doesn't actually mean that such a substance may not indeed be the endogenous ligand, but it makes the story so far pretty difficult, and I am sure there has to be a lot more research before the right compound is found.

**Lingjærde:** What about the protein found by Costa and co-workers?

**Möhler:** Well, the protein found by Costa wouldn't serve as a low molecular weight ligand. According to his theory, it is somehow situated in the membrane between the GABA receptor and the benzodiazepine receptor and helps the interregulation between these two receptors. It doesn't really float around like a neurotransmitter that is bound to the benzodiazepine receptor. A lot of labs are looking for a low molecular weight ligand like a peptide which can rapidly diffuse in the synaptic cleft.

**Marks:** Can I join in this discussion between Dr Möhler and Professor Oswald on where the benzodiazepines may be working? The fact that receptors have been shown in any particular site does not necessarily imply that those neurons are working under any circumstances. May I suggest that, in fact, the electrical activity information which is available may be more accurate in determining which neurons are working in any particular circumstance than the finding of the receptors in the brain area.

**Möhler:** I think that is quite right, and a very pertinent comment, and as far as electrophysiology goes one can only quote the different systems which have been experimentally investigated by electrophysiologists, like presynaptic inhibition of the spinal cord and so on. All the systems which have been looked at showed a similar result, namely enhancement of GABA-ergic transmission. So that whenever these neurons function physiologically, benzodiazepines probably enhance their action.

**Lingjærde:** I was very intrigued by your finding of reduced binding in Huntington's chorea. As far as I understand, you think that there are a reduced number of binding sites in Huntington's chorea. But could it also have been due to an increased concentration of the endogenous ligand?

**Möhler:** It is very difficult to exclude that, since nobody has identified the ligand. If one assumes that it is a low molecular weight ligand, then we should have washed it out during our membrane preparation, because we normally wash out all kinds of neurotransmitters and small molecular weight ligands. So I don't think we would have trouble in this respect, but it is theoretically a good point.

**Pletscher:** I was interested in this regional distribution of blood flow in the brain and metabolic activity. Is anything known about the regulation of this local blood flow and metabolism? There is some evidence that the locus caeruleus may be involved in brain microcirculation. What is your opinion about this local regulation of blood flow on metabolic activity?

**Ingvar:** Professor Pletscher, this is the 64 000 dollar question of brain physiology at present. Indeed, we do not know, as I indicated, what is the exact mechanism. The latest view – discussed at the last cerebral blood flow symposium in Copenhagen last year – was that there are probably many mechanisms which interact and which are interwoven in a way which is not well understood as yet. I would refer here again to the work carried out in Dr Siesjö's laboratory at Lund University. There is some evidence for the action of hydrogen ions. This is the old idea of Roy and Sherrington, introduced in the 1890s, that high activity makes the tissue acid, acid makes the vessels dilate, and vice versa, you have alkalinity with low activity. There is also evidence that the active nerve cells are depolarized, their membranes leak potassium ions, potassium ions affect the vessel diameters. Perhaps calcium plays a role too, and in fact there are different roles for both potassium and calcium at various pH values. Adenosine may play a role according to some workers, but not according to others. Finally, to come back to what I said to Professor Oswald, the question of the vasomotor nerves is still enigmatic. We have the nerves, they have been excellently demonstrated by workers at home in my city, Lund, by the Falk, Hillap, Uvner, Edwinson group; they have given us fantastic pictures of these nerves, and they are both catecholaminergic and cholinergic, but we still do not know what these nerves do. Probably they do not play a role in the direct coupling of function and flow, which was Professor Pletscher's question. Maybe they play a role in some general events like sleep, but this is again not known as yet. They may have a protective effect against the pressure variations which take place in emotional reactions, during physical effort, and so on, protecting the brain from the so-called breakthrough effect, but it is unlikely that there are regional projections of such nerves which could be made responsible for the coupling of function and flow. For example, if I look with my eyes to right and left, right and left, I know that my frontal eye field comes into play, a sort of light turns on here, and this probably cannot be caused by vasomotor nerve projections, but is most likely due to local factors. This question is one of the most fundamental in the whole of physiology: we do not know what makes the blood flow in the muscles increase when we move our muscles.

**Möhler:** Can one distinguish disease states in which this direct coupling between function and flow is impaired, compared with others, where the actual function is impaired and not the coupling?

**Ingvar:** Yes, I'll give you two examples of when this uncoupling takes place. The first is cerebral ischaemia. Cerebral ischaemia causes lactic acidosis in the tissue and destroys all regulation, destroys the coupling between flow and function. You may have a very high flow and a very low metabolic activity. This is the so-called luxury perfusion syndrome, a term coined by Nillsson. Some uncoupling also takes place during various forms of anaesthesia. Halothane, for example, gives rise to vasodilatation with a low activity. But we do not think that such states characterize mental disorders. It is mainly the gross organic disorders like cerebral ischaemia, stroke or brain tumour pressing upon the tissue that destroy this coupling.

**Priest:** I am very sorry to have to bring this discussion to an end. I think you will agree that we have had a stimulating morning. We have had five excellent papers, followed by a very active, perceptive discussion. I feel that we have been made aware of several growing points in sleep research and we have had lots of hints of rapid advances in knowledge soon to come.

# Section II

## Treatment of Sleep Disorders
## 1. Parameters of Efficacy

Moderator: Professor J. Bastiaans

# 6

## Clinical Pharmacokinetic and Biopharmaceutical Aspects of Hypnotic Drug Therapy

### D. D. Breimer

This chapter will deal with pharmacokinetic and biopharmaceutical aspects of hypnotic drug therapy. It will attempt to link the importance of drug elimination and drug distribution with *duration* of drug action and the importance of biopharmaceutical factors with the *onset* of drug action. Firstly some basic requirements which a hypnotic drug should fulfil in this respect will be defined.

### 6.1 THE DURATION OF DRUG ACTION

Most hypnotics are prescribed for and used by non-hospitalized or non-bedridden patients. It is therefore important that CNS-depressant effects of the drug should have declined sufficiently to be subjectively and objectively unimportant on the morning following the night of drug intake. Residual impairment of performance is undesirable, because many persons will be involved in skilled activity during day-time. So, duration of drug action should, in principle, be limited and the pharmacokinetics of a particular hypnotic and especially its distribution and elimination behaviour become important factors in achieving this[9].

Generally speaking one can say that it is an advantage for a hypnotic drug to have a relatively rapid rate of elimination (short elimination half-life). However, it does not necessarily follow that a long half-life of a drug implies a long duration of action *following single dose administration*. This point will be developed further, some theoretical and practical examples will be given and in the second part of this chapter the pharmacokinetics of some benzodiazepines will be discussed.

Hypnotics have their locus of action in the brain at certain receptor sites. This means that the intensity and duration of effect will depend on the concentration of the substances in the brain at these sites. This concentration in turn will depend on the concentration time-course elsewhere in the body, as is schematically shown in Figure 6.1. Here we are dealing with a rather complicated

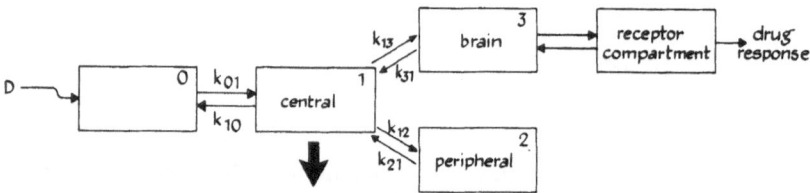

**Figure 6.1** In this theoretical schema a representation is given of the various rate processes taking place in the body after administration of a drug. The *k*-symbols represent the rate or clearance constants to and from the various compartments. The heavy arrow from the central compartment indicates the elimination process of the drug from the body, either by excretion or by biotransformation or by a combination of both

pharmacokinetic model, illustrating the exchange of a drug between various body compartments. Compartment 0 may, for instance, represent the gastrointestinal tract after oral administration of a drug, and the central compartment the plasma, blood cells, and possibly well-perfused organs and tissues. Additionally there is a peripheral compartment and a brain compartment which includes the receptors. The heavy arrow from the central compartment indicates the elimination process of the drug via the kidneys or via biotransformation for instance in the liver. This model on paper looks rather impressive and maybe even realistic, but we have to realize that in practice the possibility of determining drug concentrations will generally be limited to the central or plasma compartment. In other words we are generally only able, at least in humans,

64

to measure plasma concentrations and from these it is often rather difficult or impossible to extrapolate to the absolute concentration in certain discrete areas of the body. With respect to the concentration *time-course* in the various parts of the body it may however be somewhat simpler, if one assumes that after a certain time a distribution equilibrium is established between the various compartments, so that the concentration decay in every compartment ultimately parallels the decay in plasma.

Let us first consider the very simple situation, where the plasma concentration decrease after drug administration follows monoexponential or first-order kinetics. If one encounters such a situation it must be concluded that after administration of this drug, distribution equilibrium is reached very rapidly and that the decay of the plasma concentration is determined solely by the elimination process in the liver or kidney (metabolic or renal clearance) and the volume of distribution. It means that the duration of action can now easily be established, provided that the minimal effective concentration (MEC) of the drug is known. This situation also means that if we compare for instance a number of hypnotics, for which we assume similar distribution kinetics and a similar minimal effective concentration, their respective durations of action would be determined by their elimination half-lives only. Such a theoretical situation is

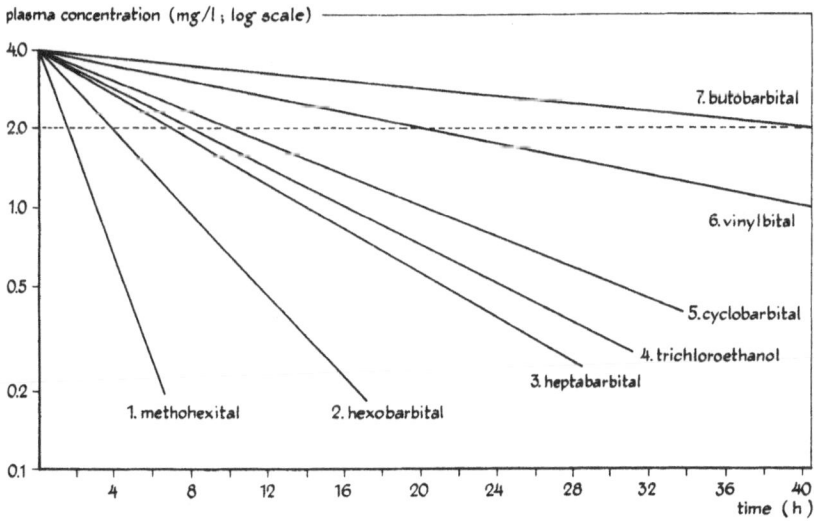

**Figure 6.2** Schematic plasma level profile of several barbiturates with different elimination half-lives (Table 6.1)

65

indicated in Figure 6.2 for various barbiturates. There are barbiturates with a relatively short half-life, e.g. hexobarbital, heptabarbital, and with a much longer half-life, e.g. butobarbital (Table
6.1)[9]. The conclusion might be that, potentially, butobarbital is
a long-acting barbiturate and hexobarbital is a short-acting one.
The traditional classification of barbiturates into long-, inter-
mediate-, short- and ultrashort-acting – as is found in most pharmacology textbooks – does suggest that their duration of action has been

**Table 6.1  Elimination half-lives of non-benzodiazepine hyp-
notics[7,9]**

| Compound | Mean elimination half-life (hours) |
| --- | --- |
| Hexobarbital | 4.0; 4.4 |
| Heptabarbital | 7.7; 9.7 |
| Cyclobarbital | 12.0 |
| Amobarbital | 20.0; 24.8; 22.7; 20.6; 23.8; 22.8 |
| Vinylbital | 23.5; 23.8 |
| Secobarbital | 23.3; 25.0; 28.9 |
| Pentobarbital | 22.3; 26.5; 29.6; 60.3 |
| Butobarbital | 37.5 |
| Glutethimide | 11.6 |
| Methaqualone | 32.6 |
| Chloral hydrate (trichloroethanol) | 8.0 |
| Ethchlorvynol | 23 |

well investigated clinically. However, this classification is based on
their duration of hypnosis or even anaesthesia when the drug is
injected intravenously into rabbits or rats and has basically nothing
to do with humans. It is surprising that this classification still
persists in pharmacology textbooks[5,26]. It is often claimed, for
example, that pentobarbital or secobarbital sodium are short-
acting compounds and hexobarbital is an intermediate-acting
agent whereas their half-lives are 20–30 hours and 4 hours respectively. As was mentioned already, very much depends on the dose
administered in relation to the MEC. But there may also be another
reason for the fact that a long elimination half-life does not necessarily imply a long duration of action. In Figure 6.3 the time–plasma
concentration profile of secobarbital after oral administration is
shown for a healthy volunteer. Absorption is rapid and after the
concentration reaches a peak there is a relatively rapid decrease

followed by a much slower decay. The more rapid phase represents mainly drug distribution from plasma into tissues. If one assumes that the drug penetrates very rapidly into the brain, in other words the brain belongs to the central compartment and thus follows the plasma concentration time-course very closely, and if one also assumes that the MEC is about 0.5 mg/l, then it becomes obvious that the duration of action is indeed limited to a few hours in this particular case. Obviously very much depends on the dose administered and on the assumption that the brain concentration parallels the plasma concentration very closely.

If differences in distribution exist, as far as penetration into or from the brain is concerned, then a more complicated picture may be obtained as is shown in Figure 6.4. Compartment 3 represents the brain compartment and 2 a peripheral one. The values for the rate constant of brain penetration have been varied and it can be seen that rapid brain penetration means continuously high concentrations (upper panel); if the rate of brain penetration equals the rate of drug elimination from the brain, then the situation as shown in the middle panel is found where there is close agreement between plasma and brain concentrations. If brain penetration is slower, then the absolute brain concentration will be lower than

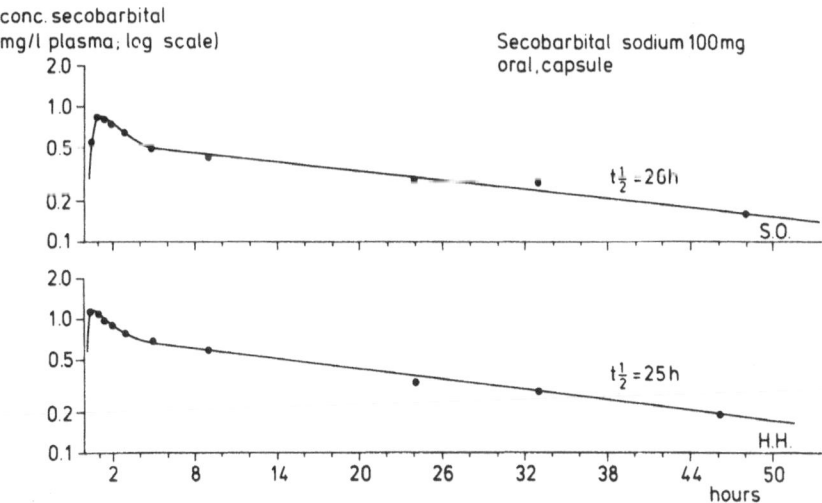

**Figure 6.3** Plasma level profile of secobarbital after oral administration of 100 mg secobarbital sodium to two healthy volunteers (reproduced from 7; permission from the author and publisher)

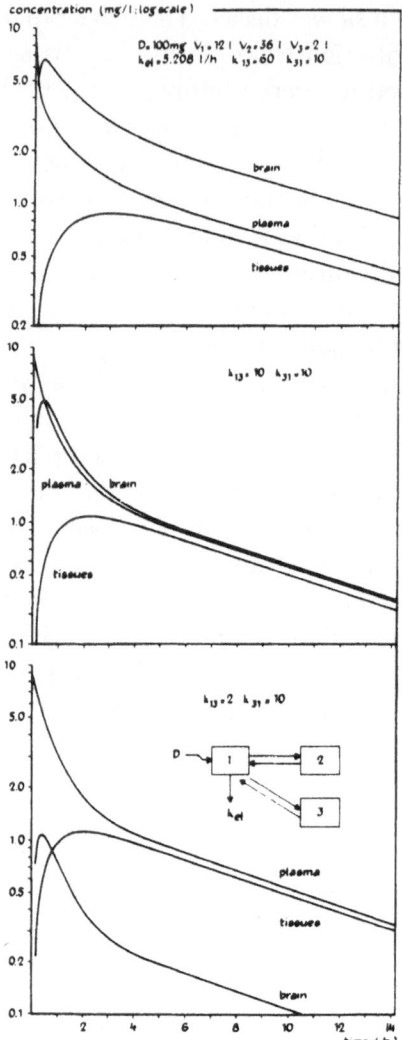

**Figure 6.4** Theoretical curves representing drug concentration in plasma and tissues rapidly equilibrating with plasma (compartment 1), the other tissues of the body (compartment 2) and the brain (compartment 3). The elimination clearance constant ($k_{el}$) is kept constant at 86.8 ml/min, whereas the clearance constant governing entry into the brain ($k_{13}$) is varied from 60 l/h to 2 l/h. *Above:* The plasma–brain clearance is 60 l/h and as a result the brain concentration rises rapidly and remains higher than the plasma concentration. *Middle:* The plasma–brain clearance equals the brain–plasma clearance ($k_{31}$) and as a result plasma and brain concentrations are practically equal. *Below:* The plasma–brain clearance is 2 l/h and now the brain concentration is always very low compared to the plasma concentration (reproduced from 33; permission from the author and the publisher)

68

plasma concentration, but still there is quite a parallelism between the time-course of brain and plasma concentrations. On the basis of these figures it seems that if a dose is given, after which concentrations are achieved higher than the MEC *only* during the distribution phase, then the duration of drug action will be limited despite a long elimination half-life. Two conditions, however, should be fulfilled for a short duration of action despite slow elimination:

1. the drug has to cross the blood–brain barrier very rapidly, so that there is instantaneous exchange of drug between plasma and brain;
2. intially the drug has to become available in the blood in relatively high concentrations, otherwise the concentration profile levels out and no distinction will be obtained between the distribution and elimination phases.

This initial high concentration can only be achieved if the drug is given by i.v. injection – which is not very convenient for general practice – or when very rapid absorption occurs when the drug is given orally. Then initial high concentrations may be obtained as is shown for secobarbital in Figure 6.3. If brain concentration closely follows plasma concentration of this hypnotic, it might be short-acting at low doses, despite its slow terminal elimination phase. Similar pictures for some benzodiazepines have been obtained, for instance for nitrazepam and flunitrazepam. Because of this rather complex situation it is not surprising that often no clear-cut relationships are found between drug effect and plasma concentration.

An important further question is: what happens if a hypnotic is taken every evening? We have already concluded earlier that there should only be effect during the night and not during the day, in other words we are pursuing an *intermittent type* of drug action when given repeatedly. This type of drug therapy is fundamentally different from most others, because then generally a constant effect at every time of day and night is desirable. In such a case the plasma concentration should be as constant as possible and constantly well above the minimal effective concentration. But for hypnotics such a situation is in fact undesirable unless one is also interested in day-time sedation, as may sometimes be the case. In Figure 6.5 it is shown what will happen with the plasma concentration for three hypnotics with different half-lives when they are given every night[9].

If heptabarbital with a half-life of 6 h is given every 24 h, no accumulation occurs and in principle one is indeed dealing with the intermittent type of drug action[10]. If on the other hand the elimination half-life of the drug is 40 h (butobarbital), then accumulation on chronic administration will occur[8]. Nitrazepam, with an average half-life of 30 h, takes an intermediate position[11]. Accumulation represents an undesirable situation, although one might argue that the patient will probably not be asleep all day despite the relatively high drug concentrations. This indicates that adaptation or tolerance to the effect of the drug develops, which may ultimately lead to drug dependence. A very undesirable situation indeed, but a condition well-known for chronic barbiturate users. If one really wants to restrict the treatment to night-time sedation and if one also wants to give the drug every night, then a short elimination half-life is certainly a great advantage and unfortunately only few hypnotics fulfil that requirement.

## 6.2 THE ONSET OF DRUG ACTION

Hypnotics are mostly prescribed for patients who experience difficulty in getting to sleep. They require a pharmaceutical formulation (dosage form) from which the active compound is rapidly absorbed. If this does not happen, early sleep may not be obtained and the patient is tempted to take a second dose, which may lead to overdosage and prolonged effect. Biopharmaceutical factors governing the rate and extent of absorption are very important in this respect, where particularly a *rapid rate of absorption* is desirable[7]. In Figure 6.6 the influence of the rate of absorption on the concentration time profile and thereby on the onset, intensity and duration of action of a drug is shown. The dose and the bioavailability of the dosage forms A, B and C are equal; thus the areas under each curve are the same. A represents a dosage form, from which the drug is so rapidly released and absorbed that a high peak concentration is achieved, such that even the safe concentration is exceeded. With a lower dose this would be a suitable profile for a *hypnotic*. B represents a rather normal profile after oral administration, and C is so slow that there will be no effect at all. A practical example, very much resembling these theoretical curves, has been published for hexobarbital[7]. When administered as its sodium salt it is very rapidly

70

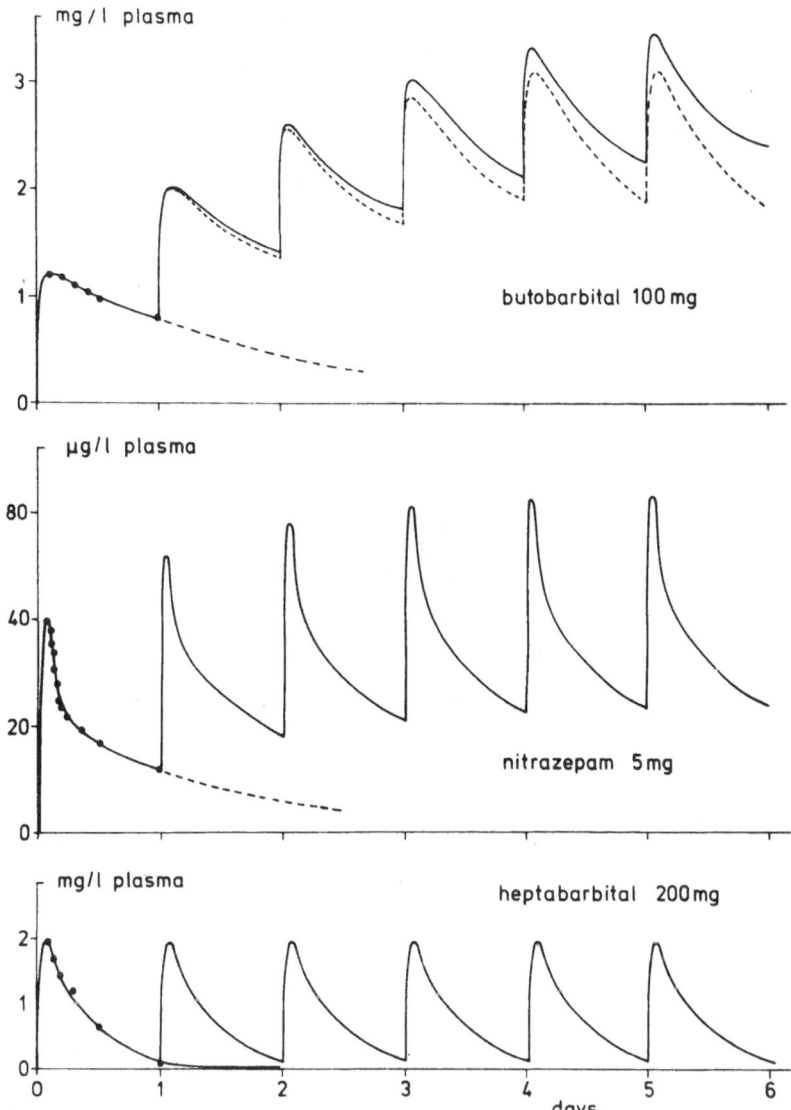

**Figure 6.5** Plasma level profile of three hypnotics during nightly administration. Calculation of the curves is based upon the experimental concentrations obtained after a single oral dose and taking into account an elimination half-life of 6 h for heptabarbital, 24 h for nitrazepam, and 40 h for butobarbital. In the case of nitrazepam the curves exhibit a biphasic decay, and for butobarbital it is shown that the elimination rate is accelerated after prolonged use owing to enzyme induction (dotted line). The figure illustrates that ideal intermittent therapy can only be achieved with a rapidly eliminated drug (heptabarbital), whereas accumulation will occur with drugs that are slowly eliminated (reproduced from 9; permission from the author and the publisher)

absorbed, resulting in high and early peak concentrations: good for rapid hypnotic onset. If given as the acid absorption takes far longer, so that a rapid onset of action cannot be expected and with a suppository formulation the situation becomes even worse.

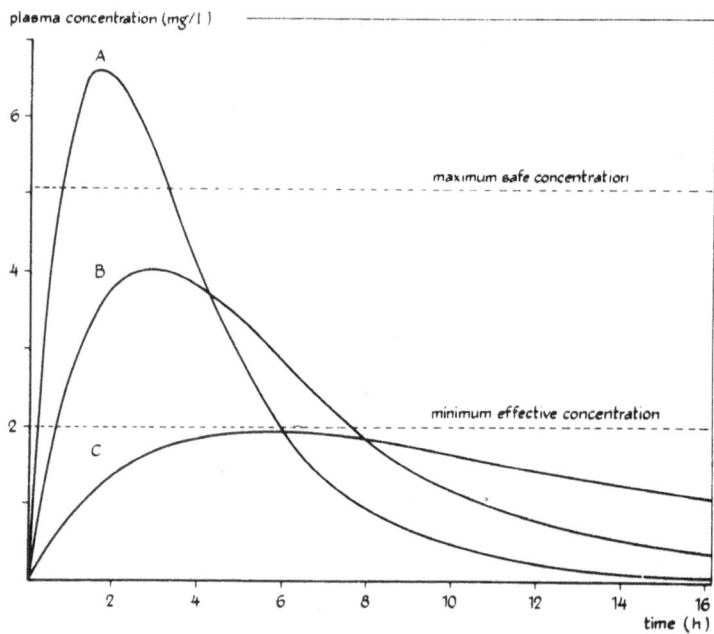

**Figure 6.6**  Theoretical plasma level profiles illustrating how changes in absorption rate may influence the onset, intensity and duration of drug action. The size of the dose and the bioavailability of the dosage forms A, B and C are the same. A: Due to the very rapid absorption the concentration exceeds the maximum safe concentration. However, by decreasing the dose this could be a suitable preparation if a very rapid onset of action is required. B: This profile may be regarded as normal after oral administration. C: Due to the slow absorption there will be no effect. However, by increasing the dose this could be a suitable preparation with prolonged action (reproduced from 7; permission from the author and the publisher)

The principle of rapid absorption is very important, not only with respect to the *onset* of action but also with regard to the *duration* of action. It was mentioned earlier that with a drug having a long elimination half-life, it is possible to obtain a short duration of action, provided that there is initially a relatively high concentration available; and the only way of getting that high concentration orally is by rapid absorption.

Differences in rates of absorption may be responsible for discrepancies in drug effects: for instance the Kales group published a paper on flunitrazepam efficacy as a hypnotic recently[3]. They found that the 2 mg dose was effective but 1 mg was not. This is in contrast to an earlier study which showed that 1 mg and even 0.25 mg was effective[18]. The discrepancy may very well be due to the differences in rates of absorption of the drug formulations used. This is an often neglected aspect of hypnotic therapy, but it does have major consequences: *with rapid absorption one may achieve a rapid onset of action with a relatively low dose and thereby also limit the duration of action.*

## 6.3 PHARMACOKINETICS OF BENZODIAZEPINES

The pharmacokinetics of some benzodiazepines, particularly those regularly used as hypnotics, will be reviewed briefly against the background of the points which have just been discussed.

**Table 6.2 Elimination half-lives of benzodiazepines and active metabolites**

| Generic name | Half-life | Active metabolite | Half-life | Reference |
|---|---|---|---|---|
| Nitrazepam | 18–36 h | none | | 11, 14, 22, 31 |
| Flurazepam | very short | $N$-desalkylflurazepam | 2–4 days | 23 |
| Flunitrazepam | 15–30 h | 7-amino-derivative | $23 \pm 4$ h | 2, 13, 34 |
| | | desmethyl-derivative | $31 \pm 8$ h | |
| Diazepam | 20–50 h | desmethyl-derivative | 30–60 h | |
| | | 3-hydroxy-derivative | | |
| | | (temazepam) | 4–10 h | 16 |
| | | oxazepam | 6–24 h | 32 |
| Temazepam | 4–10 h | none | | 16 |
| Oxazepam | 6–24 h | none | | 32 |
| Triazolam | 5–10 h | unknown | | 15 |
| Chlordiazepoxide | 7–14 h | desmethyl-derivative | | 4 |
| | | demoxepam | | |
| | | desmethyldiazepam | | |
| Lorazepam | 9–22 h | none | | 17 |
| Chlorazepate | | desmethyldiazepam | 30–60 h | 29 |

In Table 6.2 the elimination half-lives of various benzodiazepines are given, including those of a number of active metabolites. An important aspect of discussing benzodiazepines is the formation of active metabolites. When the pharmacokinetics of benzodiazepines in relation to their therapeutic effect are considered, the concentrations and concentration time-course of the active metabolites have always to be taken into account. The metabolites of diazepam are the most well-known and extensively studied and some of these have even been marketed as separate drugs.

Table 6.2 shows that there is considerable variation in the elimination half-lives of benzodiazepines. As was mentioned before: a long half-life does not necessarily imply a long duration of action after single dose administration, but it does mean accumulation if the dosage interval is small relative to its half-life (Figure 6.5).

A number of these compounds will be dealt with separately, with the exception of flunitrazepam, which is dealt with in Chapter 7.

### 6.3.1 Nitrazepam

Recently very sensitive and specific methods for the determination of nitrazepam in plasma, using gas chromatography with electron capture detection, have been developed[14, 20]. An example of the plasma concentration time-course of nitrazepam following a single oral dose is shown in Figure 6.7. Absorption is quite rapid, with early peak concentrations, and the subsequent profile is similar to that shown in Figure 6.3 for secobarbital. The initial rapid decrease of the concentration is most probably caused by distribution of the drug from plasma into tissues, whereas elimination occurs rather more slowly with an average elimination half-life of 28 hours[14] or 30 hours[11]. Similar half-life data were published by Rieder[31] and Kangas et al.[22]. The sudden rise and fall in the curve of Figure 6.7 is probably due to a redistribution phenomenon caused by the intercurrent intake of food[11], an observation similar to that made for diazepam[25].

The situation with chronic administration of nitrazepam – every night – is shown in Figure 6.5. There will be a certain degree of accumulation, particularly if the half-life is longer relative to the dosage interval. We measured nitrazepam concentration in nine hospitalized patients at 9.0 a.m. after they had been given 5 mg nitrazepam for several previous nights. There was a considerable interindividual variation in the concentration measured, with an average value of 72 ng/ml. In single-dose studies the concentrations at about 10 hours following drug administration are generally not higher than 25–30 ng/ml. There is substantial risk of accumulation, particularly in elderly or geriatric patients, because they have a significantly longer elimination half-life compared with younger people[21].

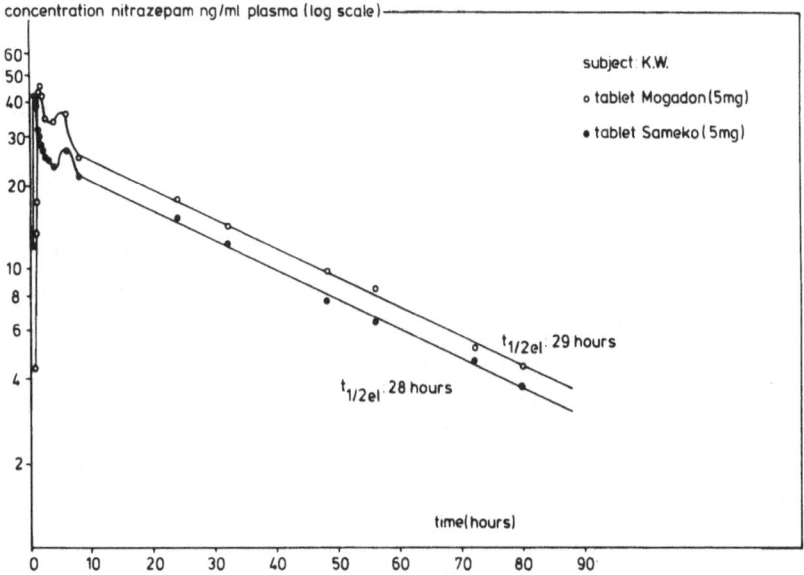

**Figure 6.7** Plasma level profile of nitrazepam after oral administration of a Mogadon tablet and a Sameko tablet, both containing 5 mg nitrazepam, to a healthy volunteer (reproduced from 14; permission from the author and the publisher)

In Holland recently quite a number of nitrazepam-containing pharmaceutical preparations have entered the market, and, as with any other drug, the question of bioequivalence – in terms of equivalent to the original Roche preparation Mogadon – becomes important. We have studied a number of these formulations, both *in vitro* and *in vivo*[12]. The *in vitro* dissolution profile showed considerable differences between the various preparations, but *in vivo* the differences were not so large. In some cases statistically significant differences with respect to the peak levels, peak level times or extent of absorption (extent of bioavailability) were found. However, since the differences were rather small, it is difficult to judge whether these would be important in practice. Unless we know more about effective concentrations, as evaluated by objective means, definite conclusions cannot be made. But bioequivalence studies are necessary in order to protect the patient against ineffective drug preparations. In principle, regulatory authorities should demand that bioequivalence is guaranteed and proven before these so-called loco-preparations enter the market[6].

### 6.3.2 Flurazepam

Kaplan et al.[23] studies the kinetics of flurazepam and its metabolites (hydroxyethyl derivative and $N$-1-desalkyl derivative) at a 30 mg daily dosage. Only trace amounts of unchanged drug were measured, whereas the hydroxyethyl metabolite was measurable for several hours after each dose. Accumulation of this metabolite was not observed. The major metabolite was the $N$-1-desalkyl derivative, which was demonstrated to possess an elimination half-life ranging from 47 to 100 hours in the four healthy volunteers. Peak concentrations after a single dose were 10–22 ng/ml blood and after 2 weeks of treatment (30 mg every night) they rose to 49–142 ng/ml. Hence, significant accumulation of this metabolite occurs during continuous administration of flurazepam.

Since this desalkylated metabolite of flurazepam has psycho-pharmacological activity similar to flurazepam in animal studies[30], it might contribute to the hypnotic and persistent sedative action of flurazepam. Kales et al.[19] have discussed the accumulation of active metabolites in relation to the effectiveness of flurazepam as evaluated in sleep laboratory studies, where peak effectiveness of the drug did not result until the second and third consecutive drug nights. It is possible indeed that one needs a relatively high concentration of this metabolite to obtain optimal hypnotic effect, but one should realize at the same time that relatively high concentrations of this active compound will be present in the body during day-time as well.

### 6.3.3 Flunitrazepam

The reader is referred to Chapter 7 on flunitrazepam by Dr Amrein[1].

### 6.3.4 Diazepam

Despite its major use as a minor tranquillizer, diazepam has been included in this brief review because there is sufficient evidence that single doses of diazepam can be applied very effectively for sleep induction[28]. In Figure 6.8 the major metabolic pathways of diazepam are shown and each of these unconjugated compounds has anti-anxiety activity. They have been marketed as single drugs (3-hydroxydiazepam or temazepam and oxazepam) or as a pro-drug (*N*-desmethyldiazepam as chlorazepate). Temazepam and oxazepam will be dealt with separately.

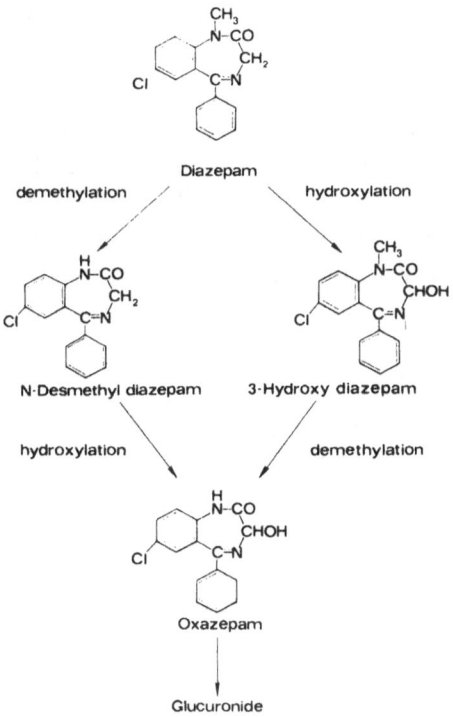

**Figure 6.8**  Metabolic pathways of diazepam in man

Numerous investigations have been published on the pharmacokinetics and metabolism of diazepam, so that no attempt will be made to refer to all the available literature. The elimination half-life of diazepam varies from 20 to 60 hours, whereas generally, in an individual subject, the half-life of desmethyldiazepam is slightly longer. There is evidence that the half-life of diazepam increases with age[24]. With chronic administration of diazepam (for instance three times daily) accumulation until relatively high steady-state levels will occur for diazepam and desmethyldiazepam.

77

If the parent drug is given for the purpose of obtaining a continuous antianxiety effect in patients, then this constant level may represent a desirable situation. When used only as a sleep-inducing agent, diazepam should not be given every night.

### 6.3.5 Temazepam (3-hydroxydiazepam)

Recently temazepam has been shown to be an effective sleep-inducing agent[27]. It has a short half-life of 5–8 hours[16] and no active metabolites are formed, except oxazepam to a very minor extent. After seven consecutive night-time administrations of temazepam to healthy volunteers, no changes in half-life became apparent and plasma concentrations were in the same order of magnitude on the seventh compared to the first day[16]. Hence from the pharmacokinetic point of view, temazepam seems to be a suitable hypnotic for administration every night.

### 6.3.6 Oxazepam

Oxazepam is predominantly used as an antianxiety agent and its pharmacokinetics have recently been reviewed[32]. It has a mean elimination half-life of 10–12 hours and it is mainly bioinactivated through glucuronidation. Although these pharmacokinetic properties are rather favourable with regard to its use as a hypnotic, its effectiveness has not been proved adequately. It is possible that its relatively polar properties prevent rapid absorption and rapid brain penetration.

### 6.3.7 Triazolam

Triazolam is a benzodiazepine derivative with a triazolo structure attached at the 3,4-position to the azepine ring. It has been shown to be an effective hypnotic in several clinical studies. Preliminary data on the pharmacokinetics and metabolism of triazolam indicate that the elimination half-life is quite short (mean of about 5 hours)[15]. Multiple dose studies (every night for 7 days) are consistent that there is not appreciable accumulation of parent compound nor metabolites. One of the major metabolites (α-hydroxytriazolam) seems to have 50–100% of the pharmacological activity of the parent compound, but this compound also appears to be rapidly eliminated[15]. More definite data on the pharmacokinetics and metabolism of triazolam are required, but so far – again just from the pharmacokinetic point of view – it seems a promising hypnotic.

### 6.3.8  Other benzodiazepines

The elimination half-lives of some other benzodiazepines which are sometimes used as hypnotics are listed in Table 6.2.

## 6.4  CONCLUDING REMARKS

In this chapter the problem of treating insomnia with hypnotics has been considered simply in terms of pharmacokinetics and biopharmaceutics, which are related to the requirements of the duration and onset of drug action. It should be realized that it is just part of the story and that many clinical and pharmacodynamic data are required to complete the whole picture for a single compound. But if we have to make a choice between hypnotics, which have comparable pharmacodynamic properties, a similar margin of safety, similar clinical effectiveness, etc., then this choice may be based upon the most appropriate pharmacokinetic properties. In addition, far more attention should be paid to the biopharmaceutical factors of hypnotic drug formulations and especially to those promoting a rapid rate of absorption. In this way, not only a rapid onset of drug action might be achieved, but at the same time a limited duration of action with a relatively low dose.

### References

1. Amrein, R., Cano, J. P., Hartmann, D., Ziegler, W. H. and Dubuis, R. (1979). Clinical and psychometric effects of flunitrazepam observed during the day in relation to pharmacokinetic data. This volume, Chapter 7
2. Amrein, R., Cano, J. P. and Hügin, W. (1976). Pharmakokinetische und pharmakodynamische Befunde nach einmaliger intravenöser, intramuskulärer und oraler Applikation von 'Rohypnol'. In Hügin, Hossli and Gemperle (eds.), *Bisherige Erfahrungen mit 'Rohypnol' (Flunitrazepam) in der Anästhesiologie und Intensivtherpie*, pp. 39–56. (Basle: Editiones Roche)
3. Bixler, E. O., Kales, A., Soldatos, C. R. and Kales, J. D. (1977). Flunitrazepam, an investigational hypnotic drug: sleep laboratory evaluations. *J. Clin. Pharmacol.*, **17**, 569

 4. Boxenbaum, H. G., Geitner, K. A., Jack, M. L., Dixon, W. R., Spiegel, H. E., Symington, J., Christian, R., Moore, J. D., Weissman, L. and Kaplan, S. A. (1977). Pharmacokinetic and biopharmaceutic profile of chlordiazepoxide HCl in healthy subjects: single-dose studies by the intravenous, intramuscular and oral routes. *J. Pharmacokinet. Biopharm.*, **5**, 3

 5. Breimer, D. D. (1975). Reclassification of barbiturates according to their pharmacokinetic behaviour in humans. Presented at the *35th International Congress of Pharmaceutical Sciences*, September 3–8, Dublin

 6. Breimer, D. D. (1976). Biologische gelijkwaardigheid van geneesmiddelen: een kwaliteitseis. *Pharm. Weekbl.*, **111**, 1121

 7. Breimer, D. D. (1976). Pharmacokinetic and biopharmaceutical aspects of hypnotic drug therapy. In Gouveia, Tognoni and van der Kleijn (eds.), *Clinical Pharmacy and Clinical Pharmacology*, pp. 17–42. (Amsterdam: Elsevier North-Holland Biomedical Press)

 8. Breimer, D. D. (1976). Pharmacokinetics of butobarbital after single and multiple oral doses in man. *Eur. J. Clin. Pharmacol.*, **10**, 263

 9. Breimer, D. D. (1977). Clinical pharmacokinetics of hypnotics. *Clin. Pharmacokinet.*, **2**, 93

10. Breimer, D. D. and de Boer, A. G. (1975). Pharmacokinetics and relative bioavailability of heptabarbital and heptabarbital sodium in man after oral administration. *Eur. J. Clin. Pharmacol.*, **9**, 169

11. Breimer, D. D., Bracht, H. and de Boer, A. G. (1977). Plasma level profile of nitrazepam following oral administration. *Br. J. Clin. Pharmacol.*, **4**, 709

12. Breimer, D. D., de Boer, A. G., Bracht, H. and Pas, J. (1978). Comparative bioavailability investigation on some nitrazepam-containing pharmaceutical preparations available on the Dutch market. (In preparation)

13. Cano, J. P., Soliva, M., Hartmann, D., Ziegler, W. H. and Amrein, R. (1977). Bioavailability of various galenic formulation of flunitrazepam. *Arzneim. Forsch.*, **27**, 2383

14. De Boer, A. G., Röst-Kaiser, J., Bracht, H. and Breimer, D. D. (1978). Assay of underivatized nitrazepam and clonazepam in plasma by capillary gas chromatography applied to pharmacokinetic and bioavailability studies in humans. *J. Chromatogr. Biomed. Appl.*, **245**, 105

15. Eberts, F. S., Ko, H. and Thomas, R. C. (1978). Metabolism and pharmacokinetics of triazolam. (In preparation)

16. Fucella, L. M., Bolcioni, G., Tamassia, V., Ferrario, L. and Tognoni, G. (1977). Human pharmacokinetics and bioavailability of temazepam administered in soft gelatin capsules. *Eur. J. Clin. Pharmacol.*, **12**, 383

17. Greenblatt, D. J., Joyce, T. H., Comer, W. H., Knowles, J. A., Shader, R. I., Kyriakopoulos, A. A., MacLaughlin, D. S. and Ruelius, H. W. (1977). Clinical pharmacokinetics of lorazepam. II. Intramuscular injection. *Clin. Pharmacol. Ther.*, **21**, 222

18. Kales, A. and Scharf, M. B. (1973). Sleep laboratory and clinical studies of the effects of benzodiazepines on sleep: flurazepam, diazepam, chlordiazepoxide and RO 5-4200. In Garrattini, Mussini and Randall (eds.), *The Benzodiazepines*, pp. 577–598. (New York: Raven Press)

19. Kales, A., Bixler, E. O., Scharf, M. and Kales, J. D. (1976). Sleep laboratory studies of flurazepam: a model for evaluating hypnotic drugs. *Clin. Pharmacol. Ther.*, **19**, 576

20. Kangas, L. (1977). Comparison of two gas-liquid chromatographic methods

for the determination of nitrazepam in plasma. *J. Chromatogr.*, **136**, 259

21. Kangas, L., Iisalo, E., Kanto, J., Lehtinen, V., Pynnönen, S., Ruikka, I., Salminen, J., Sillanpää, M. and Syvälahti, E. (1978). Human pharmacokinetics of nitrazepam: effect of age and diseases. *Eur. J. Clin. Pharmacol.* (In press)

22. Kangas, L., Kanto, J. and Syvälahti, E. (1977). Plasma nitrazepam concentrations after an acute intake and their correlation to sedation and serum growth hormone levels. *Acta Pharmacol. Toxicol.*, **41**, 74

23. Kaplan, S. A., de Silva, J. A. F., Jack, M. L., Alexander, K., Strojny, N., Weinfeld, R. E., Puglisi, C. V. and Weissman, L. (1973). Blood level profile in man following chronic oral administration of flurazepam hydrochloride. *J. Pharm. Sci.*, **62**, 1932

24. Klotz, U., Avant, G. R., Hayumpa, A., Schenker, S. and Wilkinson, G. R. (1975). The effects of age and liver disease on the disposition and elimination of diazepam in adult man. *J. Clin. Invest.*, **55**, 347

25. Kortilla, K. and Kangas, L. (1977). Unchanged protein binding and the increase of serum diazepam levels after food intake. *Acta Pharmacol. Toxicol.*, **40**, 241

26. Mark, L. C. (1969). Archaic classification of barbiturates. *Clin. Pharmacol. Ther.*, **10**, 287

27. Nicholson, A. N. and Stone, B. M. (1976). Effect of a metabolite of diazepam, 3-hydroxydiazepam (temazepam), on sleep in man. *Br. J. Clin. Pharmacol.*, **3**, 543

28. Nicholson, A. N. and Stone, B. M. (1977). Effectiveness of diazepam and its metabolite, 3-hydroxydiazepam (temazepam), for sleep during the day. *J. Physiol.*, **270**, 29P

29. Post, C., Lindgren, S., Bertler, Å. and Malmgren, H. (1977). Pharmacokinetics of N-desmethyldiazepam in healthy volunteers after single daily doses of dipotassium chlorazepate. *Psychopharmacology*, **53**, 105

30. Randall, L. O. and Kappell, B. (1973). Pharmacological activity of some benzodiazepines and their metabolites. In Garrattini, Mussini and Randall (eds.), *The Benzodiazepines*, pp. 27–51. (New York: Raven Press)

31. Rieder, J. (1973). Plasma levels and derived pharmacokinetic characteristics of unchanged nitrazepam in man. *Arzneim. Forsch.*, **23**, 212

32. Sjöqvist, F. and Sundwall, A. (eds.) (1977). The pharmacokinetic profile of oxazepam. *Acta Pharmacol. Toxicol.*, **40** (Suppl.), I

33. Van Rossum, J. M. (1973). Pharmacokinetics and psychopharmacological research. *Psychiatr. Neurol. Neurochir. (Amst.)*, **76**, 217

34. Wendt, G. (1976). Schichsal des Hypnotikums Flunitrazepam in menschlichen Organismus. In Hügin, Hossli, and Gemperle (eds.), *Bisherige Erfahrungen mit 'Rohypnol' (Flunitrazepam) in der Anästhesiologie und Intensivtherapie*, pp. 27–38. (Basle: Editiones Roche)

# 7

## Clinical and Psychometric Effects of Flunitrazepam Observed during the day in Relation to Pharmacokinetic Data

### R. Amrein, J. P. Cano, D. Hartmann, W. H. Ziegler and R. Dubuis

### 7.1 INTRODUCTION AND AIM OF THE INVESTIGATION

It may seem strange to test a benzodiazepine such as flunitrazepam, which is used mainly to induce sleep, on normal subjects during the daytime. Nevertheless, this arrangement, which was adopted for practical reasons, has been shown to have certain advantages.

The purpose of this investigation was to further clarify the pharmacokinetics of Rohypnol in a cross-over study in healthy volunteers. We were especially interested in the question of whether plasma concentrations of flunitrazepam (Figure 7.1) in the therapeutic range are strictly proportional to the dose administered. Also we wanted to observe as closely as possible the clinical effect of Rohypnol on these subjects during the experiment. Each subject therefore received doses of 1, 2 and 4 mg by mouth and 2 mg i.v. The interval between individual tests was a fortnight.

Blood samples were taken before administration of the drug and 5, 10, 15, 20, 30, 45 minutes, 1 hour, $1\frac{1}{2}$, 2, 3, 4, 5, 6, 8, 12, 24, 48, 52

## Table 7.1  Self-rating questionnaire

SELBSTBEURTEILUNG DES PATIENTEN

NAME:_____  DATUM:_____

PRÄPARAT:_____  TAGESZEIT:_____

Im folgenden finden Sie jeweils zwei extreme Fragestellungen, die sich auf Ihr gegenwärtiges Befinden beziehen. Bitte, machen Sie auf der vorgegebenen Linie dort ein Kreuz (|—X——| oder |———X—|), wo sich Ihr Befinden am ehesten zwischen den beiden Extremen einordnen lässt. Es gibt dabei keine richtigen oder falschen Antworten, es ist auch nicht wichtig, wo Sie vorher schon einmal Ihr Kreuz gesetzt haben. Bitte kreuzen Sie zügig und ohne lange zu überlegen an, und lassen Sie keine Zeile aus.

| Ich bin vollkommen ausgeruht | SR1: SEDATION | Ich möchte unbedingt schlafen |
|---|---|---|
| Ich glaube, ich habe wackelige Beine | SR2: MUSCLE STRENGTH | Ich bin voller Kraft |
| Ich weiss nichts mehr vom letzten Test | SR3: SHORT-TERM MEMORY | Ich kann mich noch ganz genau an den letzten Test erinnern |
| Ich bin entspannt | SR4: TENSION | Ich bin gespannt |
| Ich kann mich nicht konzentrieren | SR5: CONCENTRATION | Ich kann mich sehr gut konzentrieren |
| Ich fühle mich prima | SR6: GENERAL FEELING | Ich fühle mich gar nicht gut |
| Ich möchte ruhig liegen bleiben | SR7: ACTIVITY | Ich möchte aufstehen und etwas tun |
| Ich bin schwermütig | SR8: MOOD | Ich bin froh |
| Ich möchte mich mit jemandem unterhalten | SR9: SOCIABILITY | Ich möchte allein sein |
| Ich bin schlecht gelaunt | SR10: SPIRITS (MOOD) | Ich bin gut gelaunt |

DAS LETZTE BLATT TRUG IN DER OBEREN ECKE EINEN _____ PUNKT UND DIE ZAHL_____

and 58 hours after to determine the concentration of flunitrazepam (the active ingredient of Rohypnol).

Clinical and psychological tests were performed immediately after blood sampling at the following times: before ingestion of the drug, 30 minutes, 1 hour, $1\frac{1}{2}$, 2, 3, 4, 5, 6, 8, 10, 12 and 24 hours after. At each of these times, the subjects also reported their subjective condition on a ten-item bipolar self-rating scale (Table 7.1). The ten questions were designed to provide information on the degree of sedation, possible muscle-relaxing effect, concentration and memory as well as the general feeling and mood at the time.

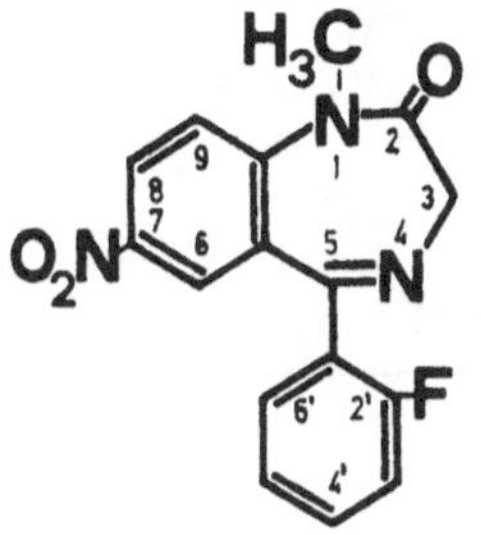

**Figure 7.1**  5-($o$-Fluorophenyl)-1,3-dihydro-1-methyl-7-nitro-2H-1,4-benzo-diazepin-2-one (flunitrazepam)

Attention and learning were tested by repeating lists of numbers forwards and backwards. Memory for colours, numbers and verbal messages was investigated and each subject also performed a tracing test (Figure 7.2) at every examination. In the last-mentioned test, the subject had to follow an 80 cm line with 13 sharp bends within a maximum of 40 seconds. Every deviation from the time counted as an error, and additional marks were lost for staying outside the boundary line for more than 1 cm and for exceeding the time limit. Each test period lasted 10 minutes, so that the subjects had the opportunity to go back to sleep between tests. In fact, they often had to be wakened for the tests, especially after the highest dose.

## 7.2  PHARMACOKINETIC RESULTS

The pharmacokinetics of Rohypnol are extremely complicated. The results after intravenous administration are the clearest. Figure 7.3 shows the curve for one subject after an intravenous dose of 2 mg

Rohypnol. You can see that the fall in plasma concentration is not even throughout the observation period. In the first hour after injection, the level of flunitrazepam in the plasma decreases extremely rapidly. Then, Rohypnol disappears from the blood at

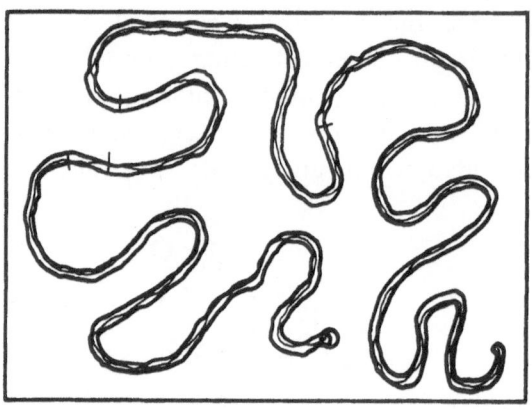

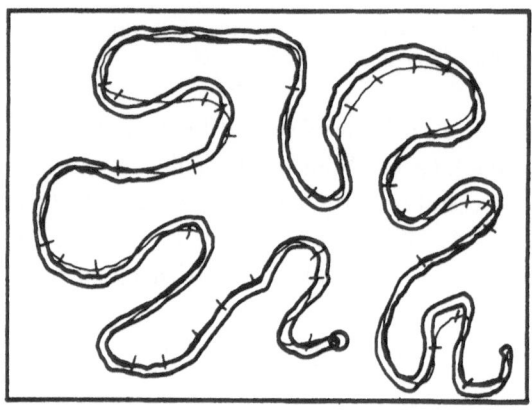

**Figure 7.2** Tracing test, two examples: Before (upper part) and 1 hour after (lower part) intake of 4 mg flunitrazepam (same patient)

a moderate rate. The last phase is reached after about 20 hours, when the plasma concentration is halved every 20 hours or so. The mathématical formula for such a curve is triexponential:

$$Cp = Ae^{-\alpha t} + Be^{-\beta t} + Ce^{-\gamma t}$$

A triexponential curve is associated with a three-compartment model, in which, in our model, elimination takes place from the central compartment (Figure 7.4). Immediately after intravenous injection, the whole amount of Rohypnol is found in the central compartment, which includes the plasma.

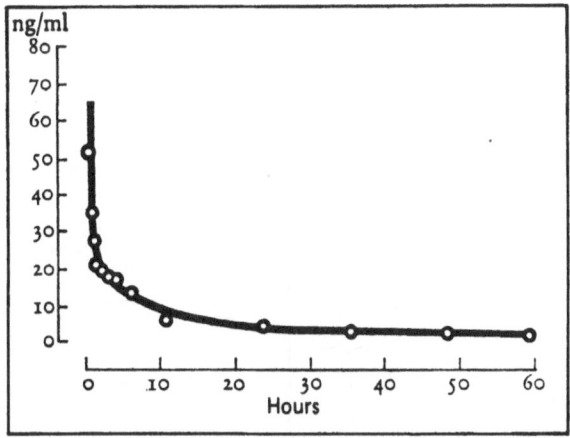

**Figure 7.3** Plasma concentration of flunitrazepam after intravenous injection of 2 mg Rohypnol

Due to rapid diffusion into the peripheral compartments, only about one-third of the amount of substance originally present remains in the first compartment after an hour (Figure 7.5), the

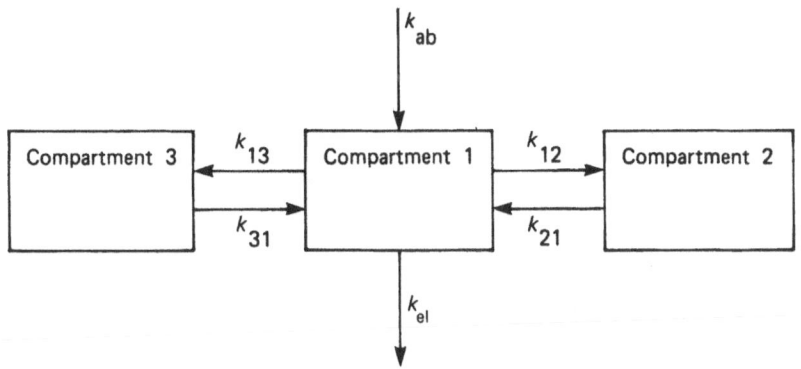

**Figure 7.4** Schematic representation of the three-compartment model used

$k_{ab}$ = absorption constant
$k_{el}$ = elimination constant
$k_{12}$ and $k_{21}$ = exchange constants between compartments 1 and 2
$k_{13}$ and $k_{31}$ = exchange constants between compartments 1 and 3

second compartment has roughly reached an equilibrium of concentration with the first and now contains about half the amount of Rohypnol injected. The third compartment contains over 14% of the Rohypnol. In succeeding hours, the concentrations in the three

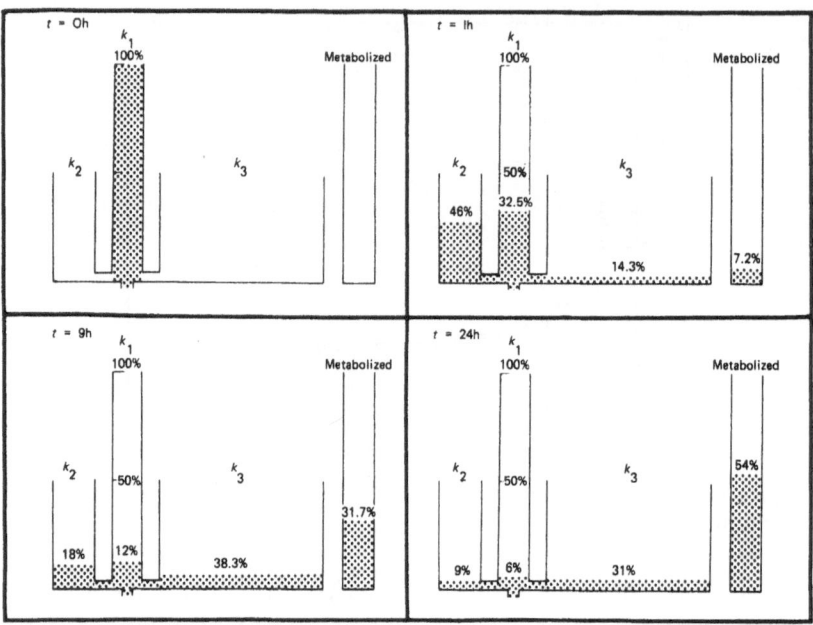

**Figure 7.5** Schematic representation of the relative amounts of flunitrazepam present in the three compartments ($k_1$, $k_2$, $k_3$) at various times. The plasma is included in the central compartment $k_1$

compartments become increasingly similar. Ten hours after administration, less than 10% of the substance remains in the central compartment. After oral administration, Rohypnol is extremely rapidly absorbed. Bioavailability of the commercial tablets is 80% or more, which means that over 90% is absorbed, approximately 10% not being available (Figure 7.6).

## 7.3 CLINICAL RESULTS

Two scales assessed the degree of sedation or alertness, two more the question of muscle relaxation and a third pair memory and ability to concentrate. The remaining four questions referred to effects on mood and feelings. As can be seen, the responses to self-rating are clearly related to dosage and time after ingestion (Figures 7.7, 7.9

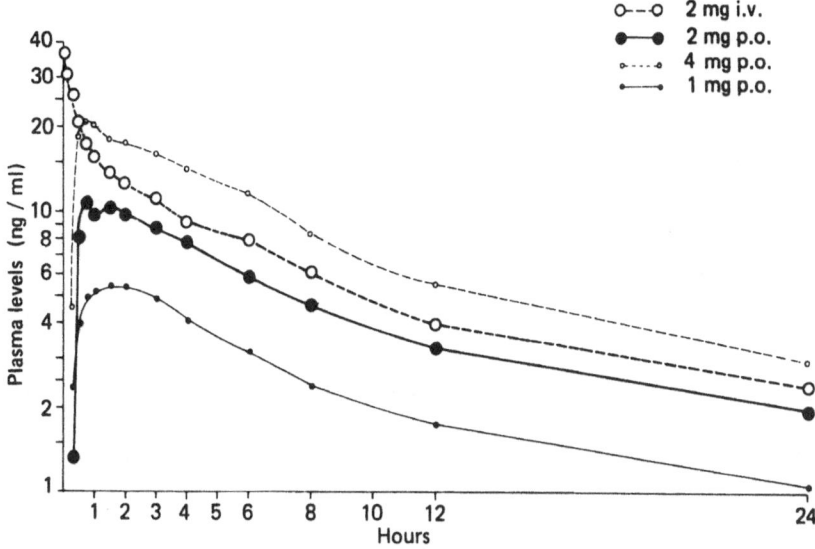

**Figure 7.6** Geometric means of plasma levels after intake of 1, 2 and 4 mg Rohypnol

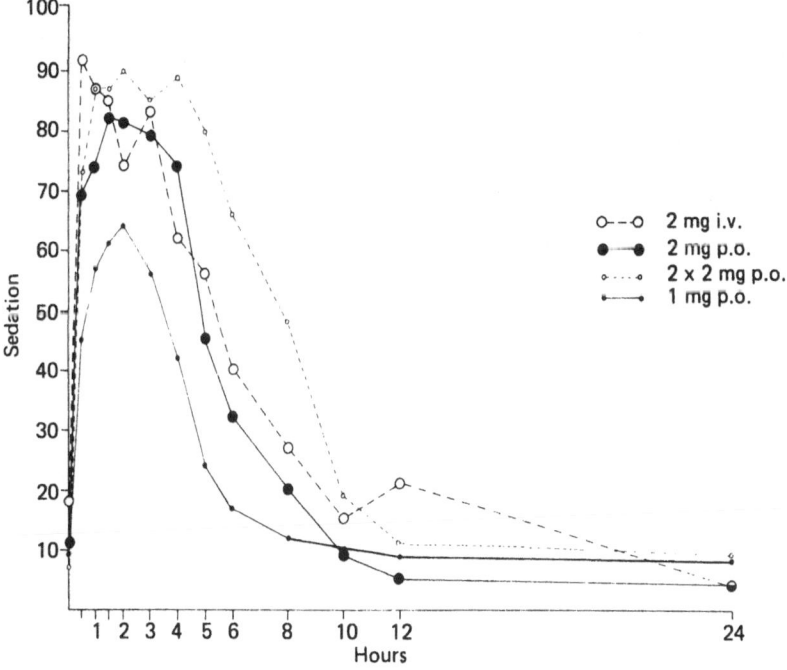

**Figure 7.7** Sedation depending on dose and time (measured in points on a 0–100 scale)

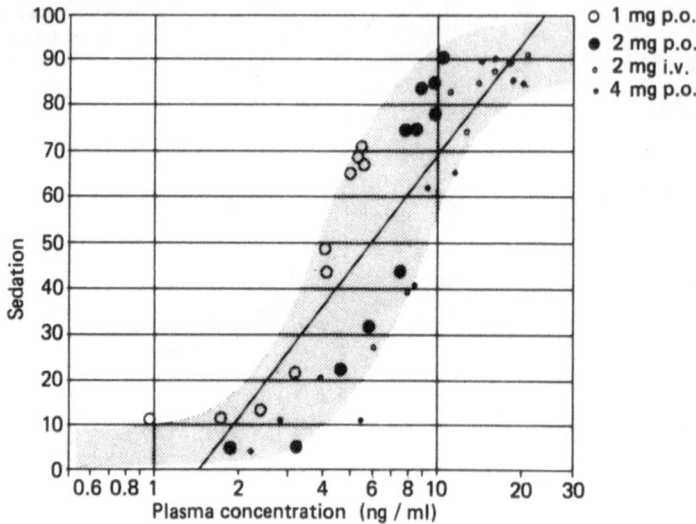

**Figure 7.8**  Correlation between plasma concentration and sedation (measured in points on a 0–100 scale)

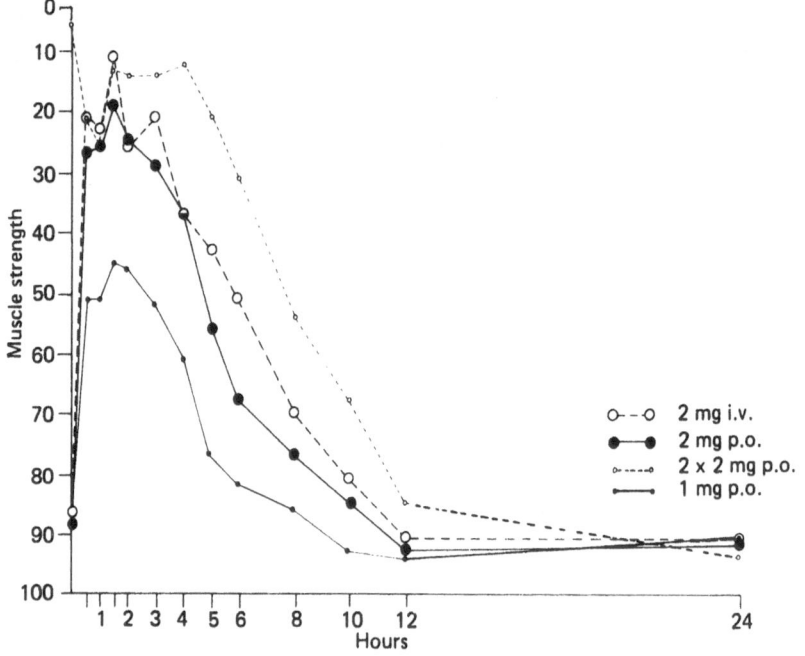

**Figure 7.9**  Muscle strength depending on dose and time (measured in points on a 0–100 scale)

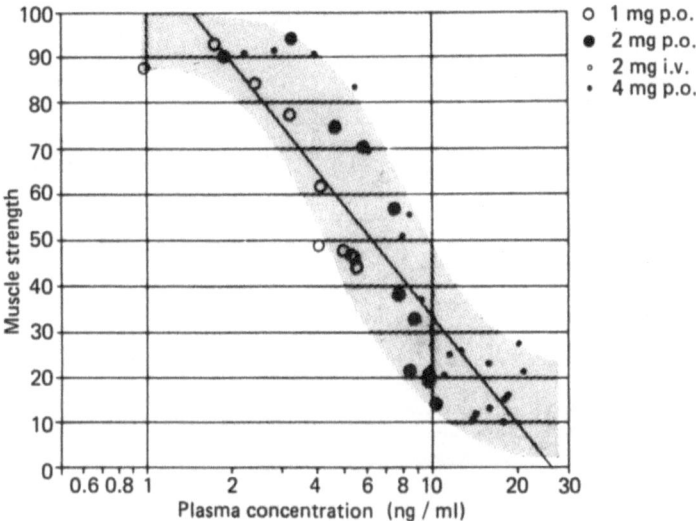

**Figure 7.10**  Correlation between plasma concentration and muscle relaxation (measured in points on a 0–100 scale)

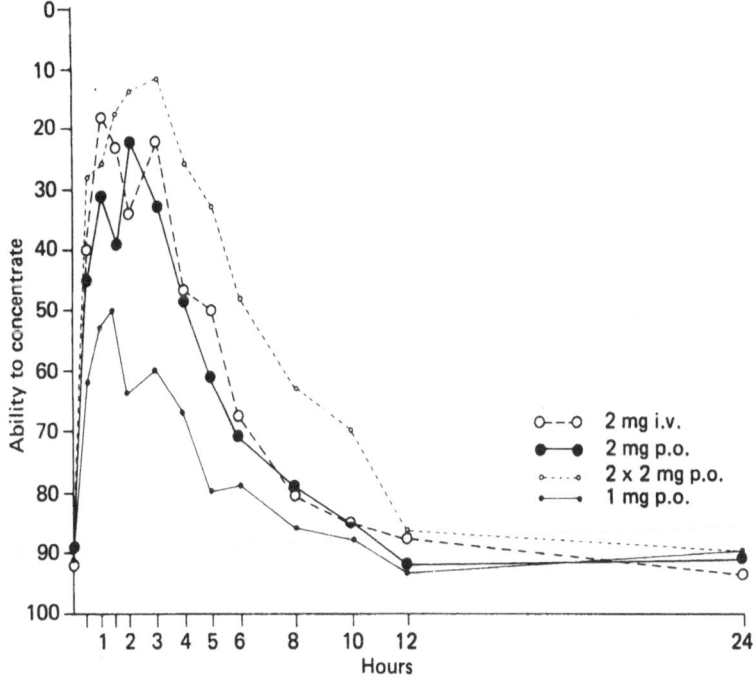

**Figure 7.11**  Ability to concentrate depending on dose and time (measured in points on a 0–100 scale)

and 7.11). Depending on the dose, the effect which subjects can recognize has disappeared after 6–12 hours. For some individuals,

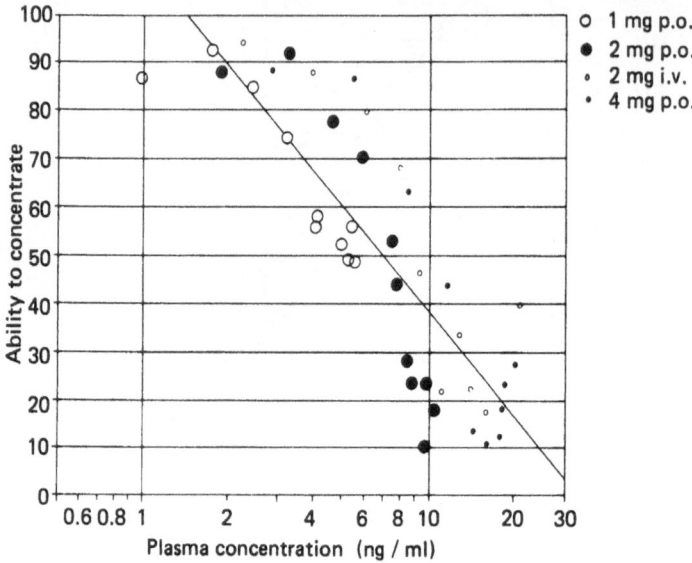

**Figure 7.12**   Correlation between plasma concentration and ability to concentrate (measured in points on a 0–100 scale)

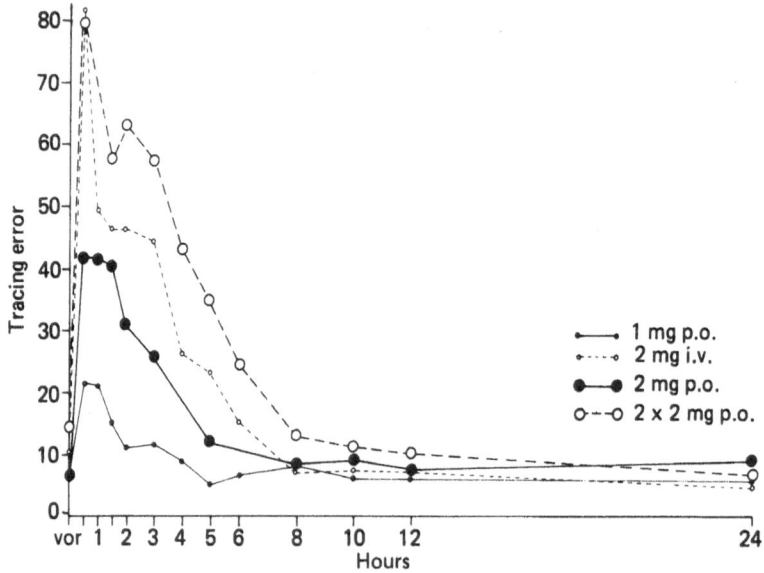

**Figure 7.13**   Errors in tracing test depending on dose and time (measured in points on a 0–100 scale)

the difference between the responses to 2 and 4 mg at the point of maximum effect is very small.

It is immediately obvious that this curve is very similar to that of the concentration of flunitrazepam in the plasma.

Various tests were performed to assess memory performance objectively. Our impression is that the various memory functions tested (memory for colours, numbers, sentences and forms) were influenced in much the same way by Rohypnol and we therefore added the individual results together in a single score for 'memory performance'. Here again the formal similarity with the plasma level curve of flunitrazepam is unmistakable.

The tracing test which we developed proved to be highly sensitive (Figure 7.13).

## 7.4 MATHEMATICAL CORRELATION BETWEEN FLUNITRAZEPAM CONCENTRATION AND PARAMETERS OF EFFECT

The individual curves show that there are quite close correlations between effect and concentration of Rohypnol.

Since Rohypnol is described by a three-compartment model, we first of all had to decide in which compartment the concentration curve most closely agreed with the clinical effect. This would enable us to allocate the effect, and therefore its substrate, to a particular compartment.

As nothing was known of the relationships to start with, we compared the clinical and psychological test results with the concentration in the different compartments by means of rank correlation. In order to calculate coefficients of rank correlation, it is not necessary to define the nature of the relationships mathematically. It is sufficient if one can assume that, in principle, higher concentrations will provoke greater effects.

In all cases, we found the highest correlation between the first compartment concentration and the clinical effect.

There was never a high correlation between the clinical effect and the concentration in the third compartment, which is hardly surprising if we realize that the high concentrations in this compartment are only reached when the clinical effect is practically finished.

For most of the parameters measured, we found a coefficient of correlation of 0.8 or higher (Table 7.2). This means that, in the present investigation, two-thirds of the clinical effect of Rohypnol

**Table 7.2 Correlation between clinical effects and pharmacokinetic parameters**

|  | Plasma concentration | Concentration first compartment | Concentration second compartment |
|---|---|---|---|
| Sedation | 0.8655 | 0.8485 | 0.7139 |
| Muscle strength | 0.8916 | 0.8809 | 0.7440 |
| Short-term memory | 0.7613 | 0.7553 | 0.8230 |
| Tension | 0.3227 | 0.3148 | 0.2744 |
| Concentration | 0.8555 | 0.8498 | 0.7284 |
| General feeling | 0.8111 | 0.7977 | 0.7291 |
| Activity | 0.8465 | 0.8316 | 0.6772 |
| Sociability | 0.8234 | 0.8051 | 0.6745 |
| Spirits (mood) | 0.8499 | 0.8360 | 0.7478 |
| Tracing test | 0.8435 | 0.8353 | 0.6653 |
| Amnesia | 0.8586 | 0.8450 | 0.7640 |

is explained by the plasma concentration and one-third is, mathematically speaking, unexplained in the sense of individual and intraindividual variations.

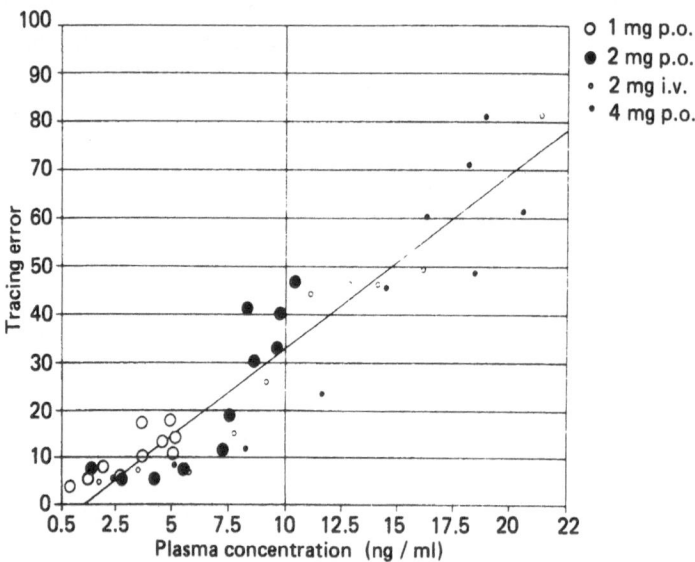

**Figure 7.14** Correlation between plasma concentration and errors in tracing test (measured in points on a 0–100 scale)

94

Theoretical considerations lead one to expect that the logarithm of the concentration would be related linearly to the clinical effect. If the clinical effect is plotted against the plasma concentration, a curve is in fact produced which primarily suggests such a relationship.

You can see that there is a fairly close correlation between the logarithm of the plasma concentration of flunitrazepam and the degree of sedation (Figure 7.8). At least in the average, central part there is a linear relationship between the logarithm concentration and the clinical effect, in this case sedation.

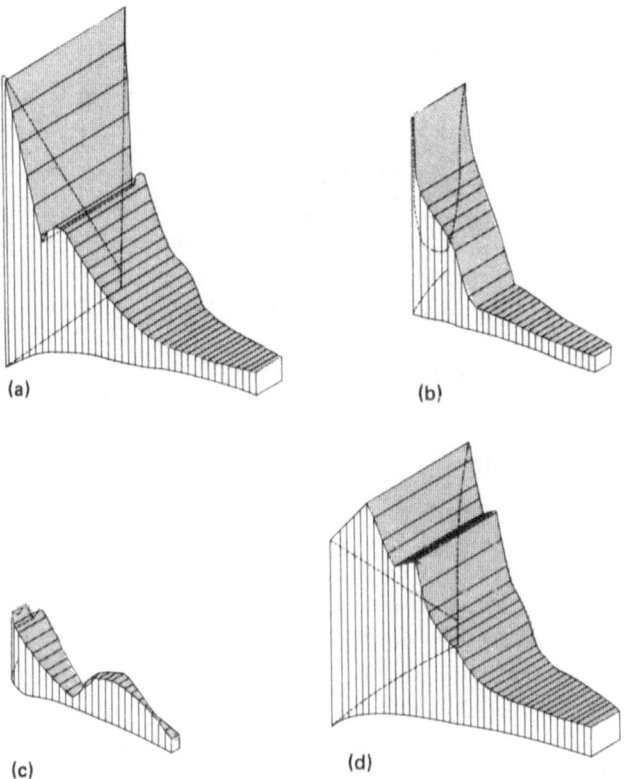

**Figure 7.15** Representation of the relationship between plasma concentration of flunitrazepam and errors in the tracing test. The number of errors is represented vertically, the plasma concentration sagittally and the time horizontally. Between the plasma concentration and the effect, there is a close correlation throughout the observation period of 12 hours.

(a) 2 mg per kg i.v.; (b) 2 mg per kg by mouth; (c) 1 mg per kg by mouth; (d) 4 mg per kg by mouth

Similar conditions apply for muscle relaxation (Figure 7.10) and also for the ability to concentrate (Figure 7.12). The amnesia score, too, can be directly related to the logarithm of the Rohypnol concentration. The same can be said also of the other self-rating scales, e.g. spirits, short-term memory, general feeling.

Among all our tests, there was only one exception, namely the tracing test (Figure 7.14). The results of this test are in linear correlation with the plasma concentration, a finding which we are unable to explain. As can be seen from Figures 7.8 and 7.10, at least some of the parameters we have examined can be described by an S-shaped regression line, with the logarithm plasma concentration of Rohypnol on the abscissa and the clinical effect observed on the ordinate. In this way, the relation between plasma concentration and clinical effect is almost ideally characterized. Of course you are familiar with curves of this kind from experimental pharmacology: they are classic dose-effect curves.

Figure 7.15 shows the changes in the tracing test, depending on the plasma concentration, that is, in relation to the dose and the time after administration. You see how close is the relationship between test result and plasma concentration even in the individual case.

## 7.5  DISCUSSION

### 7.5.1  Methodology

It seems somewhat paradoxical to test a hypnotic during the daytime. Once it had been decided, however, for the sake of uniformity, to carry out the tests on volunteers, and that it was necessary to keep waking the subjects to take blood samples and perform the clinical tests, it was obvious that the whole examination could only serve as a model.

It is known that alertness undergoes marked fluctuations over the 24 hours. The most stable conditions are to be expected between 8.0 a.m. and 7.0 p.m. This is precisely the time at which most of our experiments were conducted. Also, it seemed to us not very sensible to give the subjects a hypnotic in the evening, when they had the tendency to sleep anyway. At night-time it would be practically impossible to distinguish between the spontaneous need to sleep and that induced by the drug. For the insomniac, 'night is as the day', so that a trial of a hypnotic on healthy subjects during the daytime seems closer to the situation of the person suffering from insomnia than the same experiment at night. We therefore decided to conduct this study in the daytime, not only because in this way we avoided almost insoluble problems of discrimination, but also because this model situation seemed particularly appropriate to the problem.

We believe the results of this experiment can be extrapolated to the therapeutic situation with little modification. We found, for instance, that the subjects were most heavily sedated and had, at the most, partial recollection of events when the plasma concentration exceeded about 15 ng per ml.

We subsequently received the commission to work out a therapeutic scheme which would allow patients in the intensive care unit to be kept heavily sedated for a considerable time, without their being able to remember anything about it afterwards. We suggested a dosage schedule of Rohypnol designed to keep the plasma concentration continually above 15 ng per ml. The clinical trial of this project is not yet completed, but as far as we can say today, the dosage schedule based on the results of the experimental study reported here, seems to work in the desired way.

### 7.5.2   Elimination half-life and clinical effect

In the present study, we found, in agreement with previous investigations, that the elimination half-life of flunitrazepam in the beta-phase is about 20 hours. At first sight, it is confusing to compare this long elimination half-life with the relatively short duration of the sleep-inducing effect. Can a drug with a half-life of 20 hours ever be a good hypnotic? Would it not lead to massive cumulative effects? These theoretically quite justified questions seem to contradict the clinical experience that Rohypnol has been characterized by many thousands of trial nights as a reliable hypnotic which in therapeutically correct doses rarely causes hangover. Moreover, continuous use over many months has never led to either an increase or a decrease in effect.

This clinical observation is in full agreement with the pharmacokinetic study just reported. We believe we have been able to demonstrate that the sedative and hypnotic effects of Rohypnol are

97

correlated with its plasma concentration. As we have seen, in the first 24 hours after administration, the fluctuations in the plasma concentration are only to a small extent determined by elimination processes such as are reflected in the beta-phase half-life. You remember that 24 hours after intravenous injection of Rohypnol the plasma concentration has fallen about a twentieth of the initial value, despite the fact that, at this stage, only about half of the substance has been metabolized. The process of distribution into the deep compartment is more responsible for the observed changes in concentration than elimination. The beta-half-life remains an important parameter for Rohypnol, but in most cases it is a quite unsuitable value for describing the duration of its effect. If we wish to determine the duration of effect of Rohypnol by a pharmacokinetic measure, then this measure must be determined at a point when the effect is still evident, i.e. in the first 8–10 hours after the drug has been taken. As you know, the half-life is not stable in this period, as the distribution processes are not yet completed. As a relatively typical half-life of disappearance for the first 10 hours after ingestion, I would reckon about 3–4 hours. This would seem to be a quite reasonable relationship with the duration of the clinical effect, and to be in agreement with the clinical observation that the sleep-inducing effect has worn off about 6–8 hours after the drug has been taken and that there is generally at most a mild tranquillizing effect at this time.

## 7.6  SUMMARY

The investigations described show that there is a very close correlation between the plasma concentration of flunitrazepam on the one hand and clinically observable effects on the other. They also demonstrate that it can be quite worthwhile testing a hypnotic drug on healthy volunteers during the daytime. There are thus obviously substances like flunitrazepam which would be quite inadequately defined for the clinician by the mere recording of their elimination half-life. Such substances are characterized by the fact that the pure elimination phase is only reached after their clinical effect has already largely disappeared.

98

# 8

# Correlation of Plasma Concentrations of Benzodiazepines with Clinical Effects

**M. Lader**

## 8.1 INTRODUCTION

Although this book is primarily concerned with sleep and hypnotics, I have widened the scope of this chapter to include anxiolytic uses of the benzodiazepines, otherwise my paper would have been even more devoid of firm data than it is already. There is, however, some logic in dealing with anxiety and sleep disorders together. Firstly, many, but by no means all, insomniacs have associated high levels of anxiety, and amelioration of the anxiety is conducive to sleep. Secondly, and arising partly from the first point, the use of benzodiazepines in the treatment of insomnia can be roughly divided into two strategies, each with the most appropriate choice of anxiolytic based on the relative pharmacokinetics of the various members of the class. The clinician can try a direct approach using a single large dose of a benzodiazepine as an hypnotic. The pharmacokinetic profile of an appropriate drug would be such as to provide: (a) a rapid and dependable onset of action; (b) a short duration of action (6–8 hours); (c) no appreciable active metabolites to prolong the action. The 3-hydroxy derivatives, temazepam and lorazepam, are examples of benzodiazepines approximating these criteria. Alternatively, the physician can concentrate on lowering the general anxiety level in patients whose insomnia stems mainly from this cause. The appropriate drug would: (a) have a long-sustained action; (b) show little fluctuation in levels after each dose; (c) not result in withdrawal if the occasional dose was inadvertently omitted. Diazepam and clorazepate fulfil these criteria, their common metabolite, N-desmethyldiazepam, having a half-life in the plasma of 2–8 days[22].

## 8.2 ASSESSMENT OF EFFECTS

As with any drug, two sets of effects can be distinguished with respect to benzodiazepines, the wanted and the unwanted. Attempts can be made to relate these effects to plasma and other body-fluid

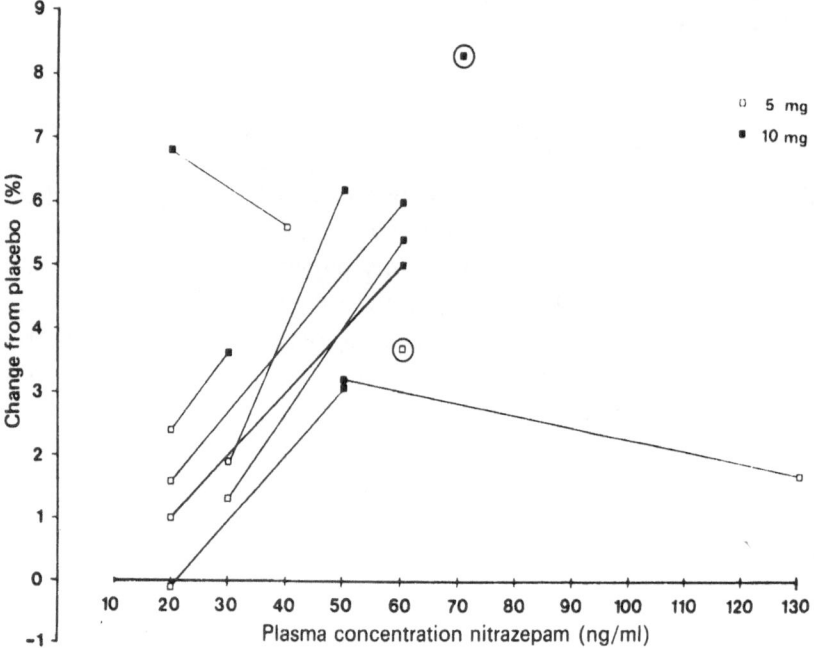

**Figure 8.1** Ten normal subjects received placebo, 5 mg and 10 mg of nitrazepam on separate occasions. The plasma concentrations of nitrazepam 12 hours after ingestion on the two drug occasions are plotted against the drug-induced increase in per cent beta activity in the EEG. The two readings from each subject are joined. The two circled points are from two subjects whose other readings were lost. Two subjects show anomalous results, but the other six show good relationships between nitrazepam levels and increase in fast-wave activity

100

drug concentrations. Immediately, with the hypnotic use of benzo-diazepines we encounter a major and possibly insurmountable obstacle. The effect that we are interested in is the sleep-inducing one which is very short-term. If we measure plasma concentrations at the point at which the patient falls asleep (assuming we overcome the practical problems with an indwelling catheter, etc.), the drug will still be in the distribution phase and absorption may still be occurring. Any plasma concentration is likely to be unrepresentative of the true drug kinetics, if not possibly misleading. Taking an estimation later during sleep to avoid these problems may be irrelevant. I shall return to this later.

The unwanted actions chiefly comprise the residual or 'hangover' effects, present the next morning. Many studies from several centres have examined this problem, usually unfortunately without plasma drug concentration estimates. In one of our studies[1], we showed some relationship between EEG changes and residual nitrazepam concentrations. Even so, as Figure 8.1 shows, anomalies occurred which vitiated any attempts at formal statistical analyses. All one can conclude is that fast-wave activity in the EEG is increased by nitrazepam roughly in proportion to its plasma concentration. No other variable was sufficiently sensitive to correlate.

Other studies have been comprehensively reviewed by Gott-schalk[8].

## 8.3 CLINICAL STUDIES

It is, however, in the clinical context that the utility of plasma benzodiazepine estimations must be examined. However, corre-lations between clinical response and plasma concentrations are weak or non-existent. Tansella et al.[21] reported no relationship between steady-state plasma concentrations of nordiazepam and hypnotic effects as measured by self or nurses' ratings. Only one of the 15 items of the Hamilton Anxiety Scale yielded a significant correlation: this was the insomnia score which dropped as drug concentrations rose. Kanto et al.[14] reported a drop in plasma con-centrations over time and no significant relationships with clinical response in 12 neurotic outpatients during subacute use or in 14 after chronic usage. In a study of flooding during waning diazepam effect,

Marks and his co-workers[16] were also unable to relate plasma concentrations to clinical effect. Our most recent studies have also been negative. In one involving 20 chronically anxious outpatients, no relationships between clinical effects and plasma concentrations were found for chlordiazepoxide, medazepam or diazepam, each given in flexible dosage for 3 weeks or more[2]. Nor was a relationship found in 24 highly anxious inpatients treated with short intensive courses of diazepam[22].

Two other studies have found some relationship. Curry[3] reported a weakly significant correlation ($r = -0.35$; $p < 0.05$) between nordiazepam concentrations and symptom ratings in nine outpatients given clorazepate. Dasberg et al.[5] found plasma concentrations of diazepam to correlate positively ($r = 0.51$; $p < 0.05$) with patient-rated main symptoms, but $N$-desmethyldiazepam concentrations to correlate negatively with autonomic symptoms ($r = -0.80$; $p < 0.01$), i.e. the patients got worse with increasing levels.

However, the general conclusion has been that it is not possible to correlate therapeutic effects of diazepam to its body concentrations[6,9]. Proposals that there is little improvement at concentrations below 400 ng per ml are still premature as there are few data to support them[13].

## 8.4  REASONS FOR THE LACK OF CORRELATION

Many causes must contribute to the failure to find any correlations of practical utility between the patient's hypnotic or anxiolytic clinical response and his plasma concentration of benzodiazepine. Clinical and pharmacokinetic causes predominate (Table 8.1).

### 8.4.1  Type of action

Drug concentrations would be expected to correlate best with clinical effects where the latter are symptom suppressants in a chronic illness. Anticonvulsants furnish a good example. Where the condition fluctuates, usually in response to external stresses and stimuli, drug control will be less satisfactory and correlations with drug levels much attenuated. The hypnotic effects are even more confusing. As outlined before, what is the clinical end-point – the point of falling asleep, the point of maximum stage 4 depth, the point of waking up, or whatever? Also it is possible that the hypnotic effects are related not to the drug levels but to their rate of increase, that is, the first differential. This has been shown for eye movements and glutethimide[4]. Such a system is too complex for routine monitoring.

**Table 8.1  Some causes of lack of correlation between plasma concentrations and clinical response**

Clinical
  Spontaneous remission
  Type of response
  Fluctuating condition
  Placebo response
Pharmacokinetic
  Flexible dosage feedback
  Pattern of metabolism
  Active metabolites
  Liver induction
  Drug interactions
  Protein binding
Technical
  Assessment of symptoms
  Quantification of response
  Measurement of drug

## 8.4.2  Metabolites

The benzodiazepines rejoice in long chains of active metabolites. Not only does each need to be measured separately but the relative potency of each with respect to the clinical effect of interest must be known. Metabolic patterns vary from patient to patient. For example, after 7 days administration of diazepam we found the ratio of diazepam to its metabolite, $N$-desmethyldiazepam, to vary from 0.60 to 2.05 (mean 1.14) among 12 patients[22]. Prior administration of amylobarbitone lowered this ratio to 0.59–1.03 (mean 0.73). Thus, concomitant or prior drug administration further confuses the issue by liver induction effects.

## 8.4.3  Induction

If a drug induces its own metabolism, complex pharmacokinetics result. The benzodiazepines induce liver microsomal enzymes in the

103

rat[12], but it is doubtful whether induction is of any clinical import-ance. However, Kanto *et al.*[14] reported that plasma diazepam con-centrations were comparatively low in patients who had taken the drug for years, whereas *N*-desmethyldiazepam concentrations were high, suggesting induction of the *N*-dealkylating enzymes. No other drugs were prescribed for these 14 patients. This conclusion has been challenged by Rutherford *et al.*[18], who found no alteration in diazepam : metabolite ratio with longer therapy. However, other medication was prescribed in this investigation which could have vitiated the results as both induction and competition could have occurred.

### 8.4.4 Plasma protein binding

The degree of plasma protein binding could influence clinical response. Benzodiazepines in clinical dosage are highly bound to plasma albumin, figures of 95–98% being commonly quoted for diazepam. Our own data[11] derived from plasma and cerebrospinal fluid estimations suggest a mean binding of at least 98.5%. The importance of this cannot be overstressed. A difference in the bound fraction of only 0.5% represents a change in the unbound, and hence biologically relevant, concentrations of diazepam of 33%, from 1.5 to 1.0. Other benzodiazepines are less highly bound[17].

### 8.4.5 Technical problems

The problems of estimating the two variables in the correlation need careful evaluation. Measuring the severity of any psychiatric con-dition is no easy task and the multitude of symptoms in anxiety and insomnia hamper the rater. Self-rating is no answer, either: each patient carries his own 'calibration rule' which makes inter-patient comparisons fraught with danger.

Nor should the chemical difficulties be overlooked. Gas–liquid chromatography, the standard method, is an involved procedure more suitable to research projects than to routine service labora-tories. Limits of sensitivity are being reached when attempts are made to estimate plasma-free concentrations.

## 8.5 FUTURE AVENUES

The inability to find meaningful correlations of clinical importance between drug concentration and clinical effect reflects in part the sporadic and rather half-hearted searches so far made. Further studies are still warranted, although investment of effort should be optimized. Some developments are worth noting.

### 8.5.1 Receptor assay

The demonstration of protein molecules in mammalian brain capable of binding benzodiazepines in proportion to their clinical potencies (allowing for pro-drugs and metabolism) has led to much research into the natural ligand, inhibiting substances and so on. It has also provided us with a simple system for assaying benzo-diazepines. Its advantages are that it assays total binding capacity of the drugs in the biological fluid, both parent and metabolites. It is quick, cheap and sensitive and could be employed on a routine service basis in any laboratories with the capability for radio-assays. Our experience with this technique has been very encouraging.

### 8.5.2 Salivary estimations

Many drugs can be measured in saliva and show good correlations with plasma total and unbound concentrations. Examples include anticonvulsants[19] and lithium[10]. Recently, diazepam has been

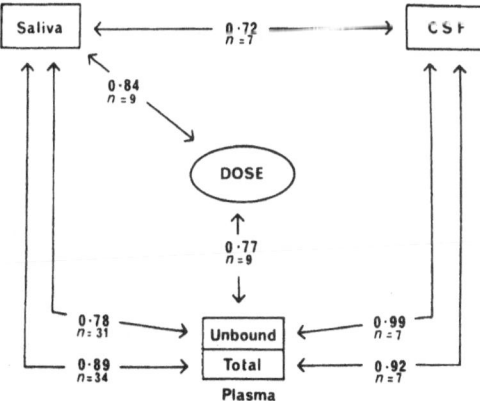

**Figure 8.2** Correlations among body fluid concentrations and dose[11]

measured in saliva after a single oral dose, but the relationship of salivary to plasma concentrations was inconstant[7]. We have confirmed this finding and extended the investigation to chronic diazepam administration (Figure 8.2)[11]. In nine chronic users, diazepam concentrations averaged 395 ng per ml, 6.5 ng per ml (1.6%) only being unbound. Salivary diazepam concentrations also averaged 6.5 ng per ml. The pooled correlation between salivary and unbound plasma concentrations was 0.78 ($p < 0.001$). Thus, salivary estimations should provide a reasonable alternative to plasma estimations, the advantage being that an estimate of unbound drug is obtained.

### 8.5.3 Relationship to dosage

Another of our findings in patients on chronic dosage was a correlation of 0.77 between daily dosage of diazepam and plasma concentration and of 0.84 with salivary diazepam levels. Such a relationship has also been reported by Zingales[23] and by Rutherford et al.[18]. The implication is that pharmacokinetic parameters do not vary much between individuals so that dosage predicts drug concentration. Therefore, variation in dosage noted, five-fold in one of our studies[15], must reflect differences in response sensitivities. This would seem to be a more heuristic approach, plasma or salivary concentration estimations being used as a monitoring variable rather than being the variable of prime interest. Response sensitivities could be assessed in a variety of ways. Test doses of diazepam could be given as a modern variant of the old Sedation Threshold Test[20]. Or EEG changes could be quantified.

## 8.6 SUMMARY

No clinically useful correlations have been reported between bodily concentrations of benzodiazepines and either hypnotic or anxiolytic effects. Reasons for this include the problem of assessing the effect induced, the presence of active metabolites, induction and drug interaction effects, plasma binding and technical limitations. Recent developments, such as simple and cheap receptor assays and the feasibility of salivary estimations, should encourage further work.

Nevertheless, it is possible that correlations with response measures may be more profitably explored.

## References

1. Bond, A. J. and Lader, M. H. (1972). Residual effects of hypnotics. *Psychopharmacologia (Berlin)*, **25**, 117–132
2. Bond, A. J., Hailey, D. M. and Lader, M. H. (1977). Plasma concentrations of benzodiazepines. *Br. J. Clin. Pharmacol.*, **4**, 51–56
3. Curry, S. H. (1974). Concentration-effect relationships with major and minor tranquillizers. *Clin. Pharmacol. Ther.*, **16**, 192–197
4. Curry, S. H. and Norris, H. (1970). Acute tolerance to a sedative in man. *Br. J. Pharmacol.*, **38**, 450–451
5. Dasberg, H. H., Van der Kleijn, E., Guelen, P. J. R. and Van Praag, H. M. (1974). Plasma concentrations of diazepam and of its metabolite *N*-desmethyldiazepam in relation to anxiolytic effect. *Clin. Pharmacol. Ther.*, **15**, 473–483
6. Garrattini, S., Marcucci, F., Morselli, P. L. and Mussini, E. (1973). In D. S. Davies and B. N. S. Prichard (eds.), *Biological Effects of Drugs in Relation to Their Plasma Concentrations*, Chapter 17, pp. 211–225. (London: Macmillan)
7. Giles, H. G., Zilm, D. H., Frecker, R. C., Macleod, S. M. and Sellers, E. M. (1977). Saliva and plasma concentrations of diazepam after a single oral dose. *Br. J. Clin. Pharmacol.*, **4**, 711–712
8. Gottschalk, L. A. (1978). Pharmacokinetics of the minor tranquillizers and clinical response. In M. A. Lipton, A. DiMascio and K. F. Killam (eds.), *Psychopharmacology: A Generation of Progress*, pp. 975–985. (New York: Raven Press)
9. Greenblatt, D. J. and Shader, R. I. (1974). *Benzodiazepines in Clinical Practice*. (New York: Raven Press)
10. Groth, U., Preilwitz, W. and Jähnchen, E. (1974). Estimation of pharmacokinetic parameters of lithium from saliva and urine. *Clin. Pharmacol. Ther.*, **16**, 490–498
11. Hallstrom, C., Curry, S. H. and Lader, M. H. (1979). Salivary concentrations of benzodiazepines. (In preparation)
12. Heubel, F. and Frank, R. (1970). Zur induktiven Wirkung von Diazepam. *Arzneim. Forsch.*, **20**, 1706–1708
13. Hillestad, L., Hansen, T. and Melsom, H. (1974). Diazepam metabolism in normal man. II. Serum concentration and clinical effect after oral administration and cumulation. *Clin. Pharmacol. Ther.*, **16**, 485–489
14. Kanto, J., Iisalo, E., Lehtinen, V. and Salminen, J. (1974). The concentrations of diazepam and its metabolites in the plasma after an acute and chronic administration. *Psychopharmacologia (Berlin)*, **36**, 123–131
15. Lader, M. H., Bond, A. J. and James, D. C. (1974). A clinical comparison of anxiolytic drug therapy. *Psychol. Med.*, **4**, 381–387
16. Marks, I. M., Visvanathan, R., Lipsedge, M. S. and Gardner, R. (1972). Enhanced relief of phobia by flooding during waning diazepam effect. *Br. J. Psychiatry*, **121**, 493–505
17. Müller, W. and Wollert, U. (1973). Characterization of the binding of benzodiazepines to human serum albumin. *Naunyn-Schmiedebergs Arch. Pharmakol.*, **280**, 229–237

18. Rutherford, D. M., Okoko, A. and Tyrer, P. J. (1978). Plasma concentrations of diazepam and desmethyldiazepam during chronic diazepam therapy. *Br. J. Clin. Pharmacol.*, **6**, 69–74
19. Schmidt, D. and Kupferberg, H. J. (1975). Diphenylhydantoin, phenobarbital, and primidone in saliva, plasma and cerebro-spinal fluid. *Epilepsia (Amsterdam)*, **16**, 735–741
20. Shagass, C. (1954). The sedation threshold. A method for estimating tension in psychiatric patients. *Electroencephalogr. Clin. Neurophysiol.*, **6**, 221–233
21. Tansella, M., Siciliani, O., Burti, L., Schiavon, M., Zimmermann-Tansella, C., Gerna, M., Tognoni, G. and Morselli, P. L. (1975). *N*-Desmethyldiazepam and amylobarbitone sodium as hypnotics in anxious patients; plasma levels, clinical efficacy and residual effects. *Psychopharmacologia (Berlin)*, **41**, 81–85
22. Tansella, M., Zimmermann-Tansella, C., Ferrario, L., Preziati, L., Tognoni, G. and Lader, M. (1978). Plasma concentrations of diazepam, nordiazepam and amylobarbitone in relation to clinical and psychological effects in anxious patients. *Pharmakopsychiatr. Neuro-Psychopharmakol.*, **11**, 68–75
23. Zingales, I. A. (1973). Diazepam metabolism during chronic medication. Unbound fraction in plasma, erythrocytes and urine. *J. Chromatogr.*, **75**, 55–78

# 9

# Effects of Benzodiazepines on Sleep and on Performance: Studies in Healthy Man

A. N. Nicholson, R. G. Borland
and B. M. Stone

## 9.1 INTRODUCTION

Within recent years there has been increasing interest in the differential effects of benzodiazepines, and many studies have been concerned with their effects on performance. The use of benzodiazepines, as both hypnotics and anxiolytics by those involved in skilled activity, has prompted the search for compounds with hypnotic activity without residual impairment and for anxiolytics with minimal impairment of performance. Our own studies have been concerned with the residual effects of drugs which would be useful in the management of disturbed sleep in those with irregular rest and activity, and these studies have involved several groups of benzodiazepines in an attempt to define compounds without residual effects on performance yet with useful hypnotic activity. Three groups of benzodiazepines have been studied. They were nordiazepam and related compounds, fosazepam and potassium clorazepate, the 1,5-benzodiazepines, clobazam and triflubazam, and diazepam and its hydroxylated metabolites, temazepam and oxazepam.

## 9.2 METHODS

Studies were carried out in healthy males and included sleep laboratory evaluations in young adulthood (20–30 years) and in middle age (45–55 years), and performance studies, in which residual as well as immediate effects were investigated with dose and time response data. Experiments were carried out in a sound-attenuated and air-conditioned room (temperature $18 \pm 1\,°C$; humidity $55 \pm 2\%$). In the performance studies the subjects were required to avoid alcohol within 24 hours of experiments and were not involved in any other form of drug therapy. There were no restrictions on the consumption of non-alcoholic beverages and they ate a light breakfast on the morning of the performance studies. In the sleep studies the subjects were familiar with the laboratory and with the techniques used in recording sleep activity. They were required to refrain from napping and undue exercise, to abstain from caffeine and alcohol after mid-day on the days which involved overnight sleep recordings, and during the whole of the day which involved day-time sleep.

Visuo-motor coordination was used to measure performance[1]. The task required the subjects to position a spot inside a randomly moving circle displayed on an oscilloscope. The movement of the spot was controlled by a hand-held stick, and an error signal proportional to the distance between the spot and the centre of the circle controlled the difficulty of the task by modulating the mean amplitude of the movement of the circle. The position of the circle and spot and the radial error signal were recorded. Subjects were trained on the task until they had reached steady performance. Each experimental run lasted 10 minutes, and during the run subjects reached their plateau performance within 100 seconds. The mean amplitude of the task over the last 500 seconds was computed, and this was the performance measure.

Sleep recordings were made with silver–silver chloride electrodes placed according to the $10:20$ system. Three electroencephalographic channels were used ($F_1$–$F_7$ or $C_4$–$A_1$; $P_1$–$T_5$ and $O_zP_z$–$O_3$), and the intervening site was selected to reduce subject discomfort and to ensure an artifact-free recording. The electromyogram was recorded from the submental musculature, and the electro-oculograms were recorded from the right eye-nasion and the left eye-nasion. Each sleep record was scored independently into 30-second epochs by two analysts. The analysis of sleep stages was carried out according to the scheme of Rechtschaffen and Kales[15]. Using the sleep-stage epochs each night's sleep was analysed for various measures[12].

The subjects also completed assessments of performance, sleep and well-being related to an analogue scale. In the sleep studies the subjects completed four assessments 30 minutes after awakening. The assessments and the extremes of the 100 mm analogue scales were: I slept *very poorly—very well*; Now I feel *very sleepy—wide awake*; I fell asleep *never—immediately*; After I fell asleep I slept *very badly— very well*. In each case a favourable response tended toward the 100 extreme of the scale.

## 9.3 RESULTS

### 9.3.1 Nordiazepam and related 1,4-benzodiazepines (Figure 9.1)

The effects of clorazepate (15 mg) and fosazepam (60 and 80 mg) on sleep were similar to the effects of their common metabolite,

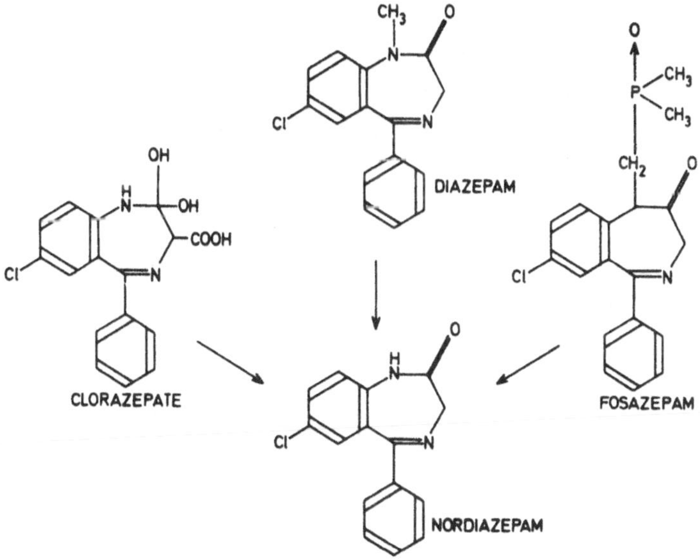

**Figure 9.1** Structural formula of nordiazepam and its precursors, clorazepate and fosazepam

nordiazepam (5 and 10 mg)[10, 12]. All three drugs increased total sleep time (Figure 9.2), hastened sleep onset and reduced awake activity and drowsy sleep (Figure 9.3). During the night after ingestion, there was reduced drowsy sleep and depressed stage 4 sleep. There were, however, some differential effects. Nordiazepam and clorazepate did not modify the latency to stage 3, though the latency was markedly shortened by fosazepam, and, though nordiazepam and fosazepam did not modify REM sleep, the first REM period was delayed with clorazepate.

With all three drugs the subjects reported improved sleep without residual impairment of well-being, and it was with these observations in mind that further studies were carried out on the immediate and residual effects of nordiazepam and its precursor, potassium clorazepate. However, somewhat surprisingly, it was difficult to establish impaired coordination during the morning after overnight ingestion of 5 and 10 mg nordiazepam or 15 mg clorazepate, though several hours after morning ingestion of clorazepate there was evidence of impaired performance (Figure 9.4). Previous

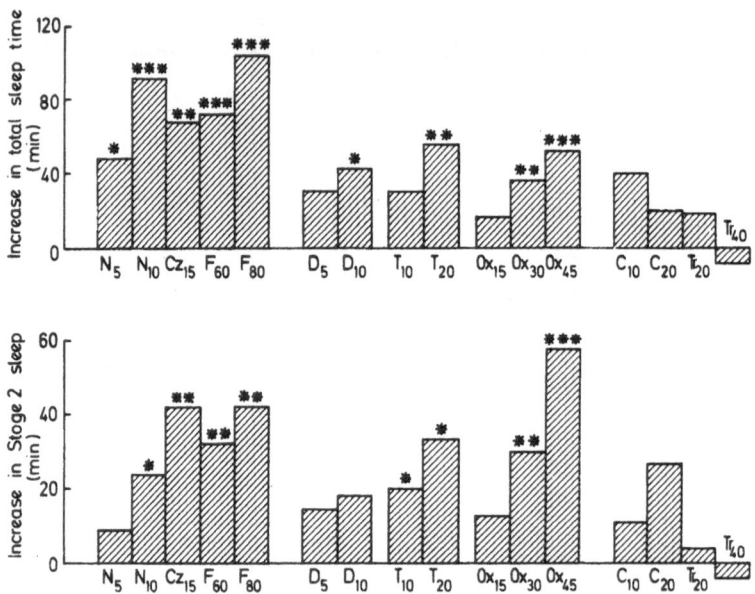

**Figure 9.2** Effects of various benzodiazepines. Increase in total sleep time and stage 2 sleep over placebo values. C clobazam, Cz clorazepate, D diazepam, F fosazepam, N nordiazepam, Ox oxazepam, T temazepam, Tr triflubazam. The figures refer to doses in mg. $*p < 0.05$; $**p < 0.01$; $***p < 0.001$

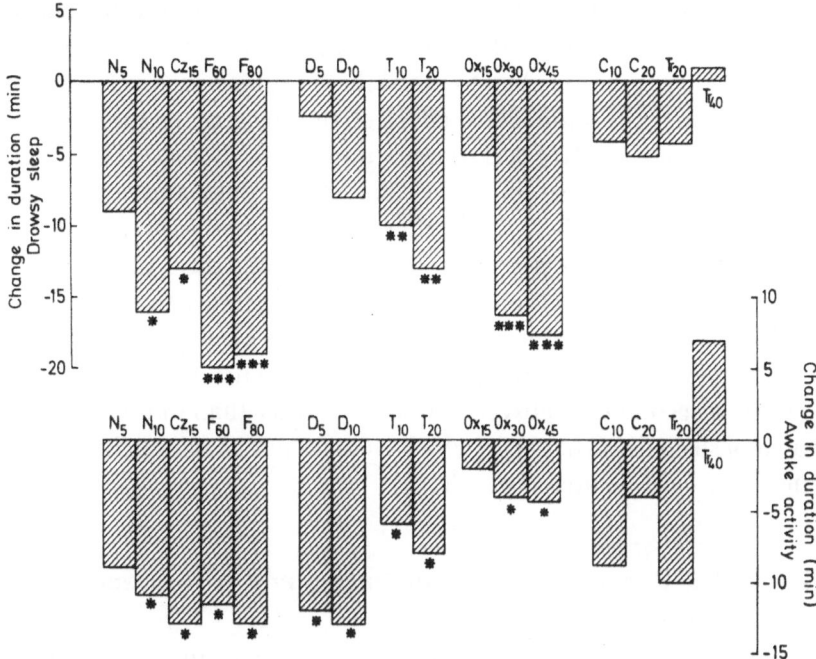

**Figure 9.3** Effects of various benzodiazepines. Decrease in awake activity and drowsy sleep over placebo values. Data for first 6 hours. C clobazam, Cz clorazepate, D diazepam, F fosazepam, N nordiazepam, Ox oxazepam, T temazepam, Tr triflubazam. The figures refer to doses in mg. $*p < 0.05$; $**p < 0.01$; $***p < 0.001$

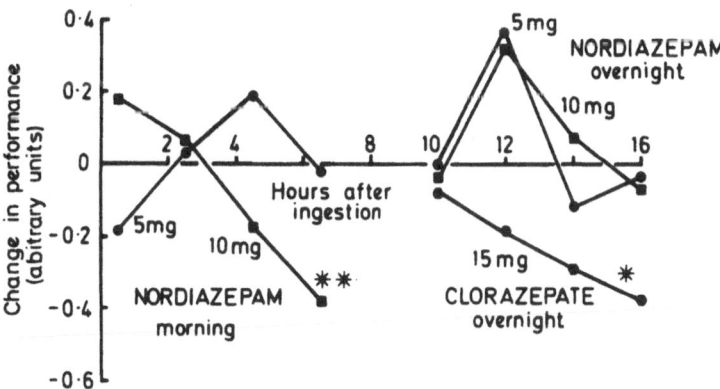

**Figure 9.4** Effect of nordiazepam (5 and 10 mg) and clorazepate (15 mg) on visuo-motor coordination. Nordiazepam was studied after morning ingestion (immediate effects) and after overnight ingestion (residual effects). Only the residual effect of clorazepate was studied. $*p < 0.01$; $**p < 0.001$

studies[14, 17] had suggested that nordiazepam may have limited effects on performance, and so the delayed appearance of impaired performance could be ascribed to the ability of subjects to overcome, at least in part, the immediate effect of the drug or to difficulty in sustaining high levels of performance under the influence of the drug.

It. was evident that nordiazepam, clorazepate and fosazepam modified sleep for 28–30 hours after ingestion, and so were likely to have anxiolytic effects during the intervening day, though with minimal residual effects on performance. The studies suggested that nordiazepam, or drugs with nordiazepam as a principal metabolite, would be more appropriate in the management of anxiety and of insomnia secondary to psychopathology rather than in the management of sleep disturbance in which changes in behaviour the following day would be unacceptable.

### 9.3.2  1,5-Benzodiazepines, clobazam and triflubazam (Figure 9.5)

With clobazam (10–20 mg) and triflubazam (20–40 mg) no effects were observed on total sleep time (Figure 9.2). Sleep onset latencies were shortened by clobazam, but this effect was not seen with triflubazam. There was reduced percentage of awake activity and drowsy sleep, but the durations of these stages were not altered (Figure 9.3). Clobazam (20 mg) reduced the percentage of stages 3 and 4, and subjects reported impaired sleep with triflubazam and less wakefulness the morning after ingestion of both drugs[11].

It appeared that the 1,5-benzodiazepines had less effect on sleep than the 1,4-benzodiazepines related to nordiazepam. However, the limited effects of clobazam could have proved useful, and so further studies were carried out on performance. The immediate effects of clobazam were studied from 0.5 to 9.5 hours after ingestion. Performance at individual times did not differ significantly from placebo (Figure 9.6), though there was some evidence of an improvement in performance during the day which could have suggested some impairment shortly after ingestion[2].

Clobazam presented as a possibly useful drug in the management of disturbed sleep when impaired performance the next day was to be avoided, except that the subjects consistently reported a sense of less wakefulness the morning after ingestion. This adverse assessment of well-being could have influenced acceptability, though difficult to reconcile with the observations on performance. Nevertheless, together with reduced stage 4 sleep, it was considered that clobazam may not have proved ideal in the management of sleep disturbance in those involved with skilled activity and was more likely to be suitable as a day-time anxiolytic.

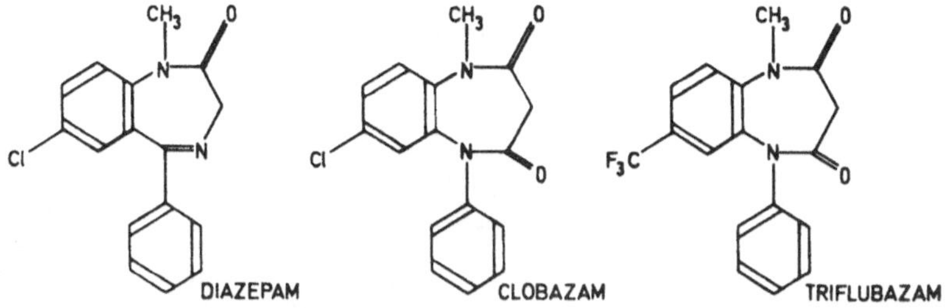

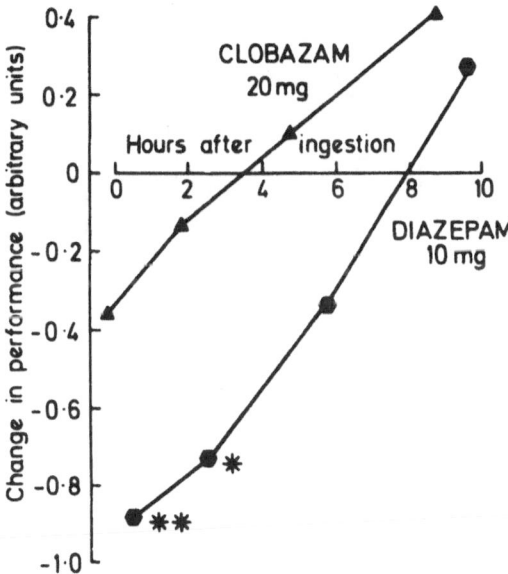

**Figure 9.5**   Structural formulae of diazepam, clobazam and triflubazam

**Figure 9.6**   Effect of clobazam (20 mg) and diazepam (10 mg) on visuo-motor coordination. Both drugs were studied after morning ingestion (immediate effects). It was impossible to establish an effect with clobazam, though a trend existed of improving performance during the day. $*p < 0.01$; $**p < 0.001$

115

### 9.3.3 Diazepam and its hydroxylated metabolites (Figure 9.7)

These studies were carried out in males aged 19–43 years[7,9,10]. Total sleep time was increased with 10 mg diazepam, sleep onset latencies were shortened and awakenings were reduced (Figures 9.2 and 9.3). With 10 mg temazepam there was no change in total sleep time, though with 20 mg total sleep time was increased. Temazepam shortened sleep onset latencies, reduced awake activity and drowsy sleep (Figures 9.2 and 9.3), and the effect on drowsy sleep was seen during each 2-hour interval of the night. The subjects reported

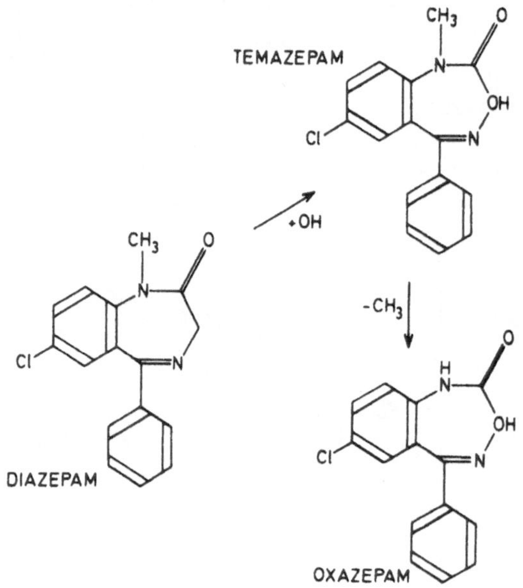

**Figure 9.7** Structural formula of diazepam and its hydroxylated metabolites, temazepam and oxazepam

improved sleep with temazepam, but subjective assessments of well-being were not altered. Oxazepam increased total sleep time, reduced awake activity and drowsy sleep (Figures 9.2 and 9.3), but it was not possible to establish an effect of this drug on sleep onset latencies. Subjects did not report changes in their quality of sleep or well-being with 15 and 30 mg oxazepam, but reported impaired wakefulness after the overnight ingestion of 45 mg. With diazepam and its metabolites there were no effects on stages 3, 4 and REM

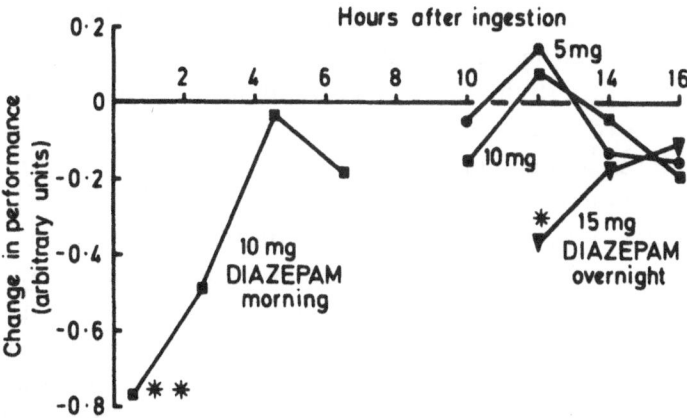

**Figure 9.8**  Effect of diazepam (5, 10 and 15 mg) on visuo-motor coordination. The effect of 10 mg diazepam was studied after morning ingestion (immediate effects) and 5, 10 and 15 mg diazepam were studied after overnight ingestion (residual effects). $*p < 0.05$; $**p < 0.01$

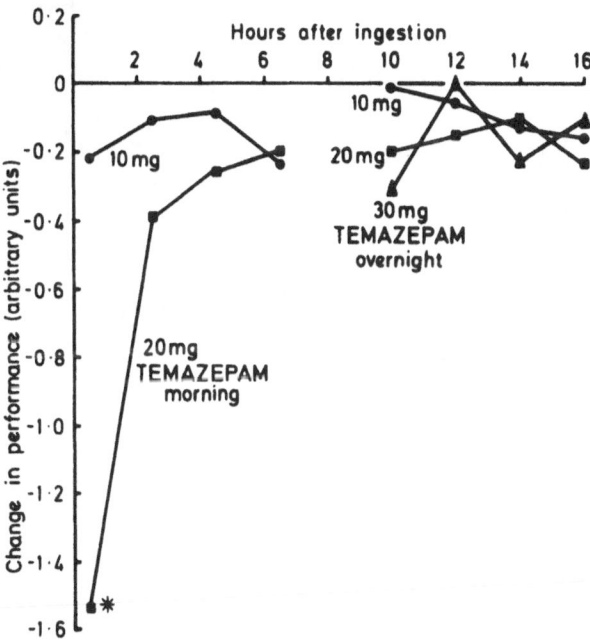

**Figure 9.9**  Effect of temazepam (10, 20 and 30 mg) on visuo-motor coordination. The effects of 10 and 20 mg temazepam were studied after morning ingestion (immediate effects) and 10, 20 and 30 mg temazepam were studied after overnight ingestion (residual effects). There was a trend toward impaired performance with 10–30 mg temazepam 10 hours after overnight ingestion. $*p < 0.001$

sleep, except that the appearance of the first REM period was delayed with 20 mg temazepam.

Performance studies were also carried out with diazepam and its hydroxylated metabolites. Performance was observed 10–16 hours after overnight ingestion of 5 and 10 mg diazepam, 10, 20 and 30 mg temazepam and 15, 30 and 45 mg oxazepam, and from 0.5 to 6.5 hours after morning ingestion of 10 mg diazepam, 20 mg temazepam and 30 mg oxazepam. Coordination was not impaired after the overnight ingestion of diazepam, temazepam or 15 and 30 mg oxazepam. There was a trend toward impaired performance with temazepam 10 hours after ingestion, and with 45 mg oxazepam performance at 10 hours was impaired (Figures 9.8, 9.9 and 9.10). With morning ingestion coordination was impaired at 0.5 and 2.5

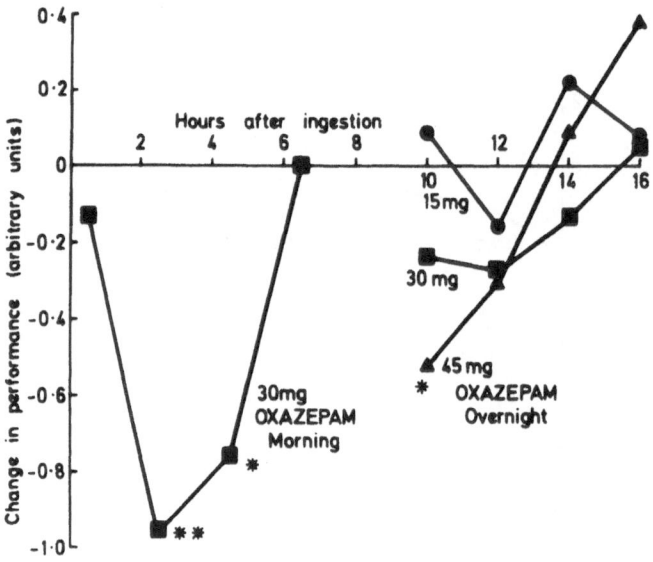

**Figure 9.10** Effect of oxazepam (15, 30 and 45 mg) on visuo-motor coordination. The effect of 30 mg oxazepam was studied after morning ingestion (immediate effect) and 15, 30 and 45 mg oxazepam were studied after overnight ingestion (residual effects). With the morning dose there was a delayed appearance of impaired performance. $*p < 0.05$; $**p < 0.001$

hours after 10 mg diazepam, at 0.5 hours after 20 mg temazepam, and at 2.5 and 4.5 hours after 30 mg oxazepam (Figure 9.11)[3].

It was evident from these and other studies[4–6, 16] that diazepam and its hydroxylated metabolites would be particularly useful in the management of disturbed sleep when impaired performance the next day was to be avoided, and it was for these reasons that further studies were carried out on the efficacy of these drugs. The effect of diazepam (5 and 10 mg) and temazepam (10, 20 and 30 mg) was studied in healthy middle age (45–55 years) males[13]. With these

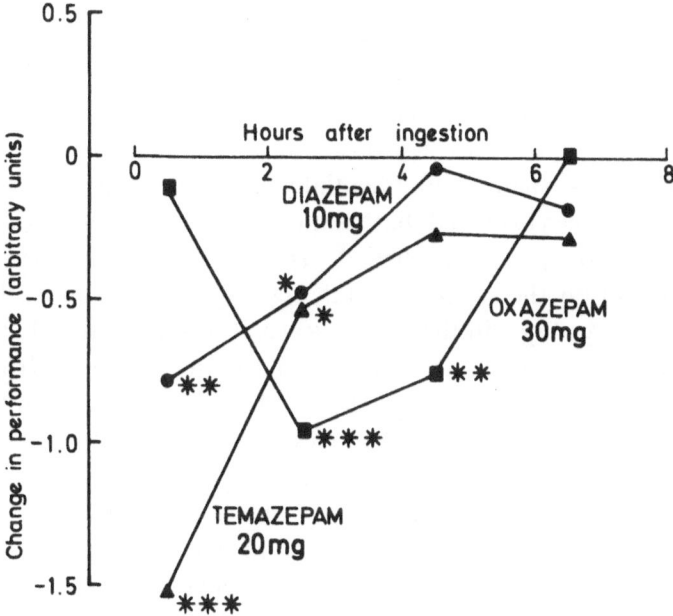

**Figure 9.11**  Immediate effects of 10 mg diazepam, 20 mg temazepam and 30 mg oxazepam on visuo-motor coordination. $*p < 0.05$; $**p < 0.01$; $***p < 0.001$

drugs there was no increase in total sleep time. Sleep onset latencies were shortened by diazepam, but were unchanged with temazepam. Awake activity was reduced by both drugs, and subjects assessed their sleep as improved with both drugs without residual effects on well-being. The effects of diazepam and its metabolites were also studied on day-time sleep from 1400 hours in young adults[8]. Total sleep time was increased by 10 and 15 mg diazepam, and by 30 and 45 mg oxazepam, but it was not possible to establish an increase in total sleep time with temazepam. Sleep onset latencies were decreased by diazepam, though such an effect was not observed with temazepam or oxazepam. Drowsy sleep was decreased by 10 and 15 mg diazepam, and by 20 mg temazepam, but there was no effect with oxazepam.

## 9.4 DISCUSSION

These studies with closely related benzodiazepines showed that very similar drugs may have different effects on sleep. In both young adulthood and in middle age diazepam reduced awake activity. The effects of temazepam and oxazepam were very similar in young adults, increasing total sleep time and reducing awake activity and drowsy sleep, though oxazepam, unlike temazepam, had no effect on sleep onset latencies. However, in middle age their effect was similar to diazepam, and essentially reduced awake activity. It would have been expected that, as the sleep of middle age is different from that of young adults[11,18], the effects of diazepam and temazepam on the sleep of middle age would have differed from that seen in young adulthood. However, it was of interest that there was no increase in total sleep time even though these drugs increased the much longer total sleep time seen in young adulthood. Further, though sleep onset latencies were similar between the two groups, and were reduced by diazepam in both groups, it was not possible to establish such an effect with temazepam which, nevertheless, shortens sleep onset latencies in early adulthood. Awake activity was reduced by each drug, but, whereas drowsy sleep was reduced by temazepam in early adulthood, there was less effect in middle age. In young adults, though the onset of the first REM period is delayed by temazepam, this could not be established in the older group.

It is also clear that the effects of diazepam and its metabolites on day-time sleep do not relate to the relative effect of the drugs on sleep during the night. Diazepam reduces awake activity and only marginally increases total sleep time at night, whereas during the day there is a marked increase in total sleep time with reduced drowsy sleep. On the other hand, temazepam and oxazepam have less hypnotic activity during the day than would be expected from night-time studies. With temazepam there is no increase in total sleep time, though there is reduced awake activity and drowsy sleep, while with oxazepam total sleep time is increased without changes in awake activity or drowsy sleep.

## 9.5 CONCLUSIONS

It is considered that diazepam and its hydroxylated metabolites provide a basis for the management of sleep disturbance in persons involved in skilled activity. In each case the drugs provide a useful hypnotic effect without residual effects on performance. Diazepam (5–10 mg) would appear to be useful for both night-time and day-time sleep, though the daily ingestion of the drug may lead to an accumulation of its long-acting metabolite with persistent behavioural effects. In this sense it may be wise to restrict ingestion to periods of not less than 48 hours and perhaps to not more than twice in a 7-day period. On the other hand, such restrictions are unlikely to be necessary with temazepam (10–20 mg) and oxazepam (15–30 mg). However, temazepam is unlikely to be an effective hypnotic for sleep at times which do not coincide with the normal circadian desire for sleep, and so may be of limited usefulness in shift workers, while oxazepam, at least in its present formulation, is unlikely to hasten sleep onset.

## 9.6 SUMMARY

The effect of three groups of benzodiazepines, nordiazepam (5 and 10 mg), and related compounds, clorazepate (15 mg) and fosazepam (60 and 80 mg), the 1,5-benzodiazepines, clobazam (10 and 20 mg) and triflubazam (20 and 40 mg), and diazepam (5, 10 and 15 mg) and its hydroxylated metabolites, temazepam (10, 20 and 30 mg) and oxazepam (15, 30 and 45 mg), have been studied on sleep and on performance in man, with particular reference to their use as hypnotics for persons involved in skilled activity.

It is considered that diazepam (5–10 mg) and its hydroxylated metabolites, temazepam (10–20 mg) and oxazepam (15–30 mg), provide a basis for the management of sleep disturbance in persons involved in skilled activity. These benzodiazepines have useful hypnotic activity without residual effects on performance.

### References

1. Borland, R. G. and Nicholson, A. N. (1974). Human performance after a barbiturate (heptabarbitone). *Br. J. Clin. Pharmacol.*, **1**, 209

2. Borland, R. G. and Nicholson, A. N. (1975). Immediate effects on human performance of a 1,5-benzodiazepine (clobazam) compared with the 1,4-benzodiazepines, chlordiazepoxide hydrochloride and diazepam. *Br. J. Clin. Pharmacol.*, **2**, 215

3. Clarke, C. H. and Nicholson, A. N. (1978). Immediate and residual effects in man of the metabolites of diazepam. *Br. J. Clin. Pharmacol.*, **6**, 325

4. Hart, J., Hill, H. M., Bye, C. E., Wilkinson, R. T. and Peck, A. W. (1976). The effects of low doses of amylobarbitone sodium and diazepam on human performance. *Br. J. Clin. Pharmacol.*, **3**, 289

5. Hindmarch, I. (1975). A 1,4-benzodiazepine: temazepam (K3917): its effect on some psychological parameters of sleep and behaviour. *Arzneim. Forsch.*, **25**, 1836

6. Molander, L. and Duvhök, C. (1976). Acute effects of oxazepam, diazepam and methylperone, alone and in combination with alcohol on sedation, co-ordination and mood. *Acta Pharmacol. (Kbh.)*. **38**, 145

7. Nicholson, A. N. and Stone, B. M. (1976). Effect of a metabolite of diazepam, 3-hydroxydiazepam (temazepam) on sleep in man. *Br. J. Clin. Pharmacol.*, **3**, 543

8. Nicholson, A. N. and Stone, B. M. (1978). Hypnotic activity during the day of diazepam and its hydroxylated metabolites, 3-hydroxydiazepam (temazepam) and 3-hydroxy, N-desmethyldiazepam (oxazepam). In A. Reinberg (ed.), *Proceedings of the Symposium on Chronopharmacology, VIIth International Congress of Pharmacology*. (Oxford: Pergamon Press)

9. Nicholson, A. N. and Stone, B. M. (1978). Hypnotic activity of 3-hydroxy, N-desmethyldiazepam (oxazepam). *Br. J. Clin. Pharmacol.*, **5**, 469

10. Nicholson, A. N., Stone, B. M. and Clarke, C. H. (1976). Effect of diazepam and a soluble derivative (fosazepam) on sleep in man. *Br. J. Clin. Pharmacol.*, **3**, 533

11. Nicholson, A. N., Stone, B. M. and Clarke, C. H. (1977). Studies on the effect of the 1,5-benzodiazepines, clobazam and triflubazam, on sleep in man. *Br. J. Clin. Pharmacol.*, **4**, 567

12. Nicholson, A. N., Stone, B. M., Clarke, C. H. and Ferres, H. M. (1976). Effect of N-desmethyldiazepam (nordiazepam) and a precursor, potassium clorazepate, on sleep in man. *Br. J. Clin. Pharmacol.*, **3**, 429

13. Nicholson, A. N. and Stone, B. M. (1979). Diazepam and 3-hydroxydiazepam (temazepam) and sleep of middle age. *Br. J. Clin. Pharmacol.* (In press)

14. Palva, E. S. and Linnoila, M. (1978). Effect of active metabolites of chlordiazepoxide and diazepam, alone or in combination with alcohol, on psychomotor skills related to driving. *Eur. J. Clin. Pharmacol.* (In press)

15. Rechtschaffen, A. and Kales, A. (1968). *A Manual of Standardized Terminology, Techniques and Scoring System for Sleep Stages of Human Subjects*. (Bethesda: United States Department of Health, Education and Welfare, Public Health Service)

16. Seppälä, T., Kortilla, K., Häkkinens, S. and Linnoila, M. (1976). Residual effects and skills related to driving after a single oral administration of diazepam, medazepam or lorazepam. *Br. J. Clin. Pharmacol.*, **3**, 831

17. Tansella, M., Zimmermann-Tansella, C. and Lader, M. (1974). The residual effects of N-desmethyldiazepam in patients. *Psychopharmacologia (Berlin)*, **38**, 81–90

18. Williams, R. L., Karacan, I. and Hursch, C. J. (1974). *Electroencephalography. EEG of Human Sleep: Clinical Applications*. (New York: Wiley & Sons)

# 10

## Amnesic Action and Residual Effects of Benzodiazepines Used for Intravenous Sedation

K. Korttila

### 10.1 INTRODUCTION

The popularity of intravenous benzodiazepines as a technique for controlling pain and anxiety during dentistry, minor surgery and diagnostic out-patient procedures has resulted largely from their capacity to induce amnesia without affecting the level of consciousness or causing depression of the cardiopulmonary system[6, 10, 34]. With such amnesic action one has come to understand a period of time after drug administration which the patients do not remember when they are later asked about it. When small doses of benzodiazepines are used, e.g. in out-patient practice, the patient may during the procedure be fully aware of the performance of the procedure but nevertheless is not able to recall the procedure afterwards. This stresses the importance of concomitant use of local anaesthetic techniques with minor sedation.

When using benzodiazepines for sedation one usually expects to achieve antianxiety and amnesic effects. However, in addition to these wanted effects the use of benzodiazepines is associated with residual effects which one would like to avoid, such as impaired psychomotor performance of patients and untoward venous sequelae in the vein used for drug injections. The purpose of this communication is to present some studies concerning the amnesic action and residual effects of three most commonly used benzodiazepines in anaesthesiological practice, diazepam, flunitrazepam and lorazepam.

## 10.2 TESTING OF AMNESIC AND RESIDUAL EFFECTS OF DRUGS

### 10.2.1 Amnesic action

Amnesic actions of drugs have commonly been evaluated by injecting these drugs to healthy volunteers or patients followed by the exposure of patients to different stimuli. Later they are asked which stimuli, e.g. visual (pictures), auditory (voices), tactile (pinching), they can recall. Similarly, one can also get an expression of the effects of dose, age or simultaneous other medication on the amnesic profile of a drug. A good example of such a technique is presented by Gregg et al.[10], when testing the amnesic action of diazepam in patients undergoing extraction of impacted third molars during local anaesthesia. With such techniques one has to remember that the anti-recall of a picture is not always the same as anti-recall of a tactile stimulus. Tactile stimuli tend to be better remembered than different pictures shown to patients[10,16].

### 10.2.2 Residual effects

In out-patient anaesthesia residual effects of drugs which impair psychomotor performance are distinct untoward effects, since these may postpone safe discharge of patients from hospitals and be deleterious outside the hospital, e.g. in traffic. When speaking of recovery from anaesthesia or sedation one should always quote the method used and distinguish clearly between immediate clinical recovery, i.e. ability to stand, walk, etc., in hospital stay, and full psychomotor recovery, i.e. ability to participate in skilful jobs or ability to drive a car. It is obvious that the more complex and sensitive tests one uses the longer one can demonstrate impaired psychomotor performance after drug administrations. Readers interested in more detailed information on the methodology used to test residual effects of drugs on psychomotor skills or to test recovery from anaesthesia or sedation are referred to pertinent reviews[14, 25].

When following venous sequelae in the vein used for drug injections one should examine the vein for more than 2 weeks, since these phenomena may not appear until 3 weeks after injections. Hewitt et al.[12] have described a method where venous complications are graded to phlebitis, thrombosis or thrombophlebitis according to the presence and extent of erythema or thrombosis of the vein or the pain felt when palpating the vein.

## 10.3 AMNESIC ACTION

### 10.3.1 Diazepam

Intravenously given diazepam induces dose-related anterograde amnesic action. When injected at a rate of 5.0 mg per minute the amnesic effect of diazepam peaks approximately $\frac{1}{2}$ to 1 minute after the cessation of injection and lasts for 5 to 15 minutes[10, 17]. Ninety per cent of healthy volunteers injected with 0.3 mg of diazepam per kg i.v. did not remember that they were pinched on the abdomen when asked afterwards[17]. When the dose was reduced to half, i.e. 0.15 mg per kg, only 30% did not remember the pinching, whereas increasing the dose to 0.45 mg per kg did not increase the amnesic effect of 0.3 mg diazepam per kg.

When diazepam was injected i.v. before the performance of bronchoscopy 59% and 30% failed to recall the performance of bronchoscopy after 0.125 mg per kg and 0.25 mg per kg, respectively, when asked during the following day[20]. Gregg et al.[10], when studying the effect of dose on the amnesic action of diazepam, suggest that increasing the dose to more than 0.3 mg per kg will result rather in the increased duration of amnesic action than in increased depth of amnesia. A rapid i.v. injection of diazepam induced greater sedative and amnesic effects than a slow injection of the same dose, but a slow injection of a greater dose is preferable because of the possibility of thrombophlebitis after rapid injection[19].

For practitioners the ptosis of the upper eyelid to cover half the pupil during i.v. injection of diazepam has been suggested as a guideline when predicting the patient to be amnesic afterwards[17, 26]. This usually corresponds to a dose of 0.3 mg diazepam per kg i.v.

without premedication in adults and less in older patients, or if simultaneous other medications, e.g. narcotic analgesics, are used[16].

### 10.3.2 Flunitrazepam

When the effect of flunitrazepam to induce amnesia for abdominal pinching was studied in young, healthy volunteers even the smallest dose of flunitrazepam used (0.01 mg per kg) caused the amnesia without affecting the level of consciousness[18]. When flunitrazepam was given i.v. before bronchoscopy only 29% and 5% of the patients remembered the procedure after 0.01 and 0.02 mg per kg, respectively[20]. These figures indicate that in dosages associated with similar side and residual effects, i.e. in a potency ratio diazepam:flunitrazepam 10:1, flunitrazepam has slightly better amnesic action than diazepam. This has also been suggested by George and Dundee[9] who noticed a slightly longer duration of amnesia after flunitrazepam than diazepam injections.

When the effect of age on flunitrazepam-induced amnesia and sedation was studied during local anaesthesia for bronchoscopy there was an increase in anti-recall of bronchoscopy with increasing age, but the most distinct differences between different age groups were that the amnesic action of flunitrazepam started earlier and persisted longer in patients over 60 than in those under 60 years of age[21]. According to our experience, when amnesia is sought after i.v. injection of flunitrazepam, following doses should induce amnesic action of 5–15 minutes duration for 80–100% of patients:

> patients under 40 years: flunitrazepam 0.02 mg/kg;
> patients 40–59 years: flunitrazepam 0.015 mg/kg;
> patients over 60 years: flunitrazepam 0.01 mg/kg.

Contrary to i.v. diazepam amnesia can be expected after i.v. flunitrazepam even if the ptosis of the upper eyelid is not sufficient to cover half the pupil[18].

### 10.3.3 Lorazepam

The amnesic action of lorazepam differs from those of diazepam and flunitrazepam in two respects: first, even after intravenous injection the onset of amnesic effect of lorazepam is slow; secondly, the duration of amnesic action of lorazepam is distinctly longer than those of other benzodiazepines.

Pandit et al.[29] showed in a well-controlled study that lorazepam 2 mg i.v. produced a short anti-recall effect in 50% of patients with a latency of 30 minutes and a duration of less than half an hour. After 4 mg lorazepam i.v. more than 70% of the patients were amnesic for visual stimuli 15 minutes to 4 hours after injection. Sedation was long-lasting following both doses of lorazepam, but was not related to the anti-recall effect[29]. Pagano et al.[28] followed thoroughly with memory cards the onset of anti-recall effect of lorazepam. They noticed that after 2 mg lorazepam i.v. only 8% and 40% of patients were amnesic 4 and 30 minutes after the injection, respectively; the respective percentages for patients given 4 mg lorazepam i.v. being 37% at 4 minutes and 70% at 30 minutes.

Lorazepam has been shown to produce distinct amnesic action also when given orally or i.m. Dundee et al.[7] gave 4 mg lorazepam orally or i.m. as premedicants to patients undergoing minor gynaecological operations. Forty per cent of patients did not remember the journey to the operating theatre and i.v. injection of the anaesthetic induction agent. With oral or i.m. route of administration the duration of amnesia after lorazepam is similar to that noticed after i.v. injection, but the onset of action is even slower with the former modes of administration.

### 10.3.4  Relative amnesic actions

When the relative amnesic actions are compared the most distinct difference is that with lorazepam the onset of amnesia is slower and lasts longer (for up to 4 hours) than is the case with diazepam and flunitrazepam. The incidence of amnesia is slightly more frequent and its duration slightly longer after flunitrazepam than after diazepam.

## 10.4  RESIDUAL EFFECTS

### 10.4.1  Psychomotor skills

#### 10.4.1.1  Diazepam and flunitrazepam

With small doses both the immediate recovery and recovery of psychomotor skills is similar after small doses of diazepam and fluni-trazepam, but if the dose is increased to more than 0.3 mg per kg versus 0.02 mg per kg of diazepam and flunitrazepam, respectively, both immediate and complete recovery are slower after flunitrazepam than after diazepam[17, 18, 20]. It took healthy volunteers an average of 36 minutes before they could stand steady after i.v. injection of 0.3 or 0.45 mg of diazepam per kg[17], but subjects injected with 0.03 mg of flunitrazepam per kg could not stand steadily before 90 minutes after injection and vertigo or unsteady gait were common still 4 hours afterwards[18]. Orr *et al.*[27] were able to demonstrate impaired standing steadiness of volunteers for $2\frac{1}{2}$ hours and 4 hours after 10 and 20 mg oral diazepam. They did not test flunitrazepam with the same method. Patients' ability to stand and walk on a straight line were similar after 0.125 mg of diazepam per kg i.v. and 0.01 mg of flunitrazepam per kg i.v., but after 0.25 mg of diazepam per kg and 0.02 mg of flunitrazepam per kg i.v. these functions normalized more slowly after flunitrazepam[20]. When the effect of age was studied on recovery after flunitrazepam sedation, the patients' eye coordination and their ability to stand steadily and walk on a line normalized more slowly in patients over 60 than in those under 60 years of age, but no differences in recovery were noted between patients under 40 and those 40–59 years of age or between those 60–69 and those over 70 years of age[21].

When the residual effects of diazepam and flunitrazepam have been tested with psychomotor test battery, coordinative skills have always been affected most and for the longest time[16 –18, 20]. After intravenous diazepam, healthy volunteers' coordinative skills were impaired for 4, 6 and 8 hours after doses of 0.15, 0.30 and 0.45 mg per kg, respectively[17]. After 0.01 mg of intravenous flunitrazepam per kg, eye–hand coordination was slightly impaired for as long as 6 hours after injection, and after 0.02 and 0.03 mg per kg the impairment was still distinct at the last observation period 10 hours after injection[18]. Bond and Lader[2] have also shown a motor impairment in behavioural tests 12 hours and altered EEG up to 18 hours after 1-mg and 2-mg doses of oral flunitrazepam. Because of slow recovery and long-lasting residual effects on psychomotor skills doses over 0.02 mg of flunitrazepam per kg should be avoided in out-patient anaesthesia or sedation.

Baird and Hailey[1] first reported that after i.v. sedation with diazepam, clinical sedation with an increase in plasma diazepam concentrations may recur at 6 hours and beyond. Later we have repeatedly shown this late elevation to be due to food intake[15,19,24], which presumably remobilizes diazepam from some storage site, e.g. from the liver or the wall of the gastrointestinal tract[19]. This elevation of serum diazepam may induce a late impairment of psychomotor skills, especially if the food is eaten within less than 5 hours after injection[19]. From a practical point of view it seems to me that, although this late elevation is true also with oral and i.m. administration of diazepam, the mode of administration should be i.v. and the dose more than 0.3 mg per kg before any clinically significant late impairment can be expected.

### 10.4.1.2  Lorazepam

Residual effects of lorazepam have not been studied as extensively as those after diazepam and flunitrazepam. After lorazepam premedication in women undergoing minor gynaecological operations the maximum sedative effect and drowsiness persisted for at least 4 hours and patients showed residual drowsiness which continued for up to 6 hours.

Stoller et al.[32], using visual tracking as a method of detecting the residual effects of lorazepam, found that impaired hand–eye coordination may persist for 4 hours after lorazepam 2 mg i.m. and 8 hours after lorazepam 4 mg i.m. Seppälä et al.[31] compared residual effects and skills related to driving after oral administration of diazepam (10 mg) and lorazepam (2.5 mg). Lorazepam impaired almost all the measured skills more than diazepam, and the lorazepam impairment of reactive skills and flicker fusion discrimination remained statistically significant for as long as 12 hours.

The long-lasting drowsiness and residual effects of psychomotor skills indicate that lorazepam should not be used in out-patient anaesthesia or sedation as stressed by many investigators[7,31,32]. Dundee et al.[7] stress that medicolegal implications of long-lasting amnesia must be also remembered, particularly if the patient is told something about the operation or given instructions under the influence of lorazepam.

### 10.4.2  Venous sequelae

Intravenous administration of diazepam has been reported to be associated with an unacceptably high incidence of thrombophlebitis[23], presumably because of diazepam precipitations penetrating into the vein[13]. Vein-irritating properties of diazepam can be minimized if care is taken not to inject the drug more rapidly than at a rate of 5 mg per minute if large veins are used for injections and if the vein is flushed with saline after diazepam injection.

Since the vein-irritating properties of diazepam have been suggested to be attributable to its solvent propylene glycol, attempts have been made to use other solvents for diazepam. Von Dardel *et al.*[5] suggest that the use of diazepam in an emulsion form is not associated with venous sequelae in the vein used as injection site. Similarly, diazepam with cremophor EL as solvent should not cause thrombophlebitis. But the use of cremophor EL for this purpose has seriously been questioned because of the hypersensitivity reactions associated with diazepam dissolved in cremophor EL[3]. One preparation with polyethylene glycol as a main solvent for diazepam has by *in vitro* studies been shown to cause less precipitation with intravenous solutions than diazepam with propylene glycol as solvent, indicating the former to be associated with less venous sequelae than with the conventional solvent[22]. However, clinical evidence is still lacking with this formula.

Comparative clinical studies on the incidence of venous sequelae after intravenous injections of different benzodiazepines have not been presented until recently by Hegarty and Dundee[11]. They report in their abstract a distinctly greater incidence of venous sequelae after i.v. diazepam (39%) than after flunitrazepam (5%) or lorazepam (15%). Thornton *et al.*[33] have also suggested flunitrazepam used in sedation for dentistry to cause less thrombophlebitis than diazepam used for the same purpose.

## 10.5 CONCLUSIONS

The present communication gives an overview of the amnesic actions and residual effects of three different benzodiazepines which should give a basis on which one could employ a specific benzodiazepine in relation to the desired rapidity of onset and duration of amnesia as well as in relation to the acceptable amount of residual effects. Diazepam induces a rapid amnesic action of short duration and may have residual effects for as long as 10 hours after injection.

Venous sequelae are rather common after i.v. diazepam. Fluni-trazepam induces a rapid amnesic action of short (intermediate) duration and may have residual effects for as long as 10–24 hours after injection, depending on dosage. Lorazepam has a slowly start-ing amnesic action of long duration and has distinct residual effects for as long as 24 hours after its injection. Lorazepam has an amnesic effect also when given orally or i.m. Flunitrazepam more than 0.02 mg per kg or lorazepam in any dosage capable of producing amnesia should be avoided in out-patient anaesthesia or sedation.

## 10.6 FUTURE

It appears that intravenous sedation has become a well-established technique for controlling pain and anxiety during various minor procedures. An ideal drug for this purpose would be a drug which does not have cardiorespiratory side-effects or other untoward effects and induces rapid and well-predictable amnesic action of short duration without having either long-lasting residual effects on psychomotor performance or any untoward venous sequelae. Recently a new water-soluble benzodiazepine, midazolam maleate (Ro 21-3981), has been suggested to have, at least to some extent, these properties[4,8,30]. Hitherto, information on this newly syn-thesized benzodiazepine is rather limited, and it is too early to predict its future usefulness in anaesthesia or sedation. Studies carried out similarly to those reviewed in this communication will reveal the future place of new drugs among other benzodiazepines now available.

## References

1. Baird, E. S. and Hailey, D. M. (1972). Delayed recovery from a sedative: correlation of the plasma levels of diazepam with clinical effects after oral and intravenous administration. *Br. J. Anaesth.,* **44**. 803
2. Bond, A. J. and Lader, M. H. (1975). Residual effects of flunitrazepam. *Br. J. Clin. Pharmacol.,* **2**, 143
3. Clarke, R. S. J. (1978). A survey of hypersensitivity reactions to intravenous anaesthetics. Symposium Abstracts on *Adverse Responses to Intravenous Agents.* Sheffield University Medical School, Sheffield, UK, July 1978
4. Conner, J. T., Katz, R. L., Pagano, R. R. and Graham, C. W. (1978). Ro 21-3981 for intravenous surgical premedication and induction of anaes-thesia. *Curr. Res. Anesth. Analg.,* **57**, 1

5. Von Dardel, O., Mebius, C. and Mossberg, T. (1976). Diazepam in emulsion form for intravenous usage. *Acta Anaesthesiol. Scand.*, **20**, 221

6. Duncan, A. W. and Barr, A. M. (1973). Diazepam premedication and awareness during general anaesthesia for bronchoscopy and laryngoscopy. *Br. J. Anaesth.*, **45**, 1150

7. Dundee, J. W., Lilburn, J. K., Nair, S. G. and George, K. A. (1977). Studies of drugs given before anaesthesia. XXVI: Lorazepam. *Br. J. Anaesth.*, **49**, 1047

8. Fragen, R. J., Gahl, F. and Caldwell, N. (1978). A water-soluble benzodiazepine, Ro 21-3981, for induction of anaesthesia. *Anesthesiology*, **49**, 41

9. George, K. A. and Dundee, J. W. (1977). Relative amnesic actions of diazepam, flunitrazepam and lorazepam in man. *Br. J. Clin. Pharmacol.*, **4**, 45

10. Gregg, J. M., Ryan, D. E. and Levin, K. H. (1974). The amnesic actions of diazepam. *J. Oral Surg.*, **32**, 651

11. Hegarty, J. E. and Dundee, J. W. (1978). Local sequelae following the i.v. injection of three benzodiazepines. *Br. J. Anaesth.*, **50**, 78

12. Hewitt, J. C., Hamilton, R. C., O'Donnel, J. F. and Dundee, J. W. (1966). Clinical studies of induction agents. XIV. A comparative study of venous complications following thiopentone, methohexitone and propanidid. *Br. J. Anaesth.*, **38**, 115

13. Jusko, W. J., Gretch, M. and Gasset, R. (1973). Precipitation of diazepam from intravenous preparations. *J. Am. Med. Assoc.*, **225**, 176

14. Korttila, K. (1976). Minor outpatient anaesthesia and driving. *Mod. Probl. Pharmacopsychiatry (Basel)*, **11**, 91

15. Korttila, K. and Kangas, L. (1977). Unchanged protein binding and the increase of serum diazepam levels after food intake. *Acta Pharmacol. Toxicol.*, **40**, 241

16. Korttila, K. and Linnoila, M. (1974). Skills related to driving after intravenous diazepam, flunitrazepam or droperidol. *Br. J. Anaesth.*, **46**, 961

17. Korttila, K. and Linnoila, M. (1975). Recovery and skills related to driving after intravenous sedation: dose–response relationship with diazepam. *Br. J. Anaesth.*, **47**, 457

18. Korttila, K. and Linnoila, M. (1976). Amnesic action and skills related to driving after intravenous flunitrazepam. *Acta Anaesthesiol. Scand.*, **20**, 160

19. Korttila, K., Mattila, M. J. and Linnoila, M. (1976). Prolonged recovery after diazepam sedation: the influence of food, charcoal ingestion and injection rate on the effects of intravenous diazepam. *Br. J. Anaesth.*, **48**, 333

20. Korttila, K., Saarnivaara, L., Tarkkanen, J., *et al.* (1978). Comparison of diazepam and flunitrazepam for sedation during local anaesthesia for bronchoscopy. *Br. J. Anaesth.*, **50**, 281

21. Korttila, K., Saarnivaara, L., Tarkkanen, J., *et al.* (1978). Effect of age on flunitrazepam-induced amnesia and sedation during local anaesthesia for bronchoscopy. *Br. J. Anaesth.* (In press)

22. Korttila, K., Sothman, A. and Andersson, P. (1976). Polyethylene glycol as a solvent for diazepam: bioavailability and clinical effects after intramuscular administration, comparison of oral, intramuscular and rectal administration, and precipitation from intravenous solutions. *Acta Pharmacol. (Kbh)*, **39**, 104

23. Langdon, D. E., Harlan, M. J. R. and Bailey, R. L. (1973). Thrombophlebitis with diazepam used intravenously. *J. Am. Med. Assoc.*, **223**, 184

24. Linnoila, M., Korttila, K. and Mattila, M. J. (1975). Effect of food and

repeated injections on diazepam serum levels. *Acta Pharmacol. (Kbh)*, **36**, 181

25. Linnoila, M., Saario, I., Seppälä, T., *et al.* (1974). Methods used for evaluation of the combined effects of alcohol and drugs on humans. In P. L. Morselli, S. Garattini and P. J. Cohen (eds.), *Drug Interactions*, p. 319. (New York: Raven Press)
26. O'Neil, R., Verril, P. J., Aellig, W. H. and Laurence, D. R. (1970). Intravenous diazepam in minor oral surgery. *Br. Dent. J.*, **127**, 15
27. Orr, J., Dussault, P., Chappel, C., Goldberg, L. and Reggiani, G. (1976). Relation between drug-induced central nervous system effects and plasma levels of diazepam in man. *Mod. Probl. Pharmacopsychiatry (Basel)*, **11**, 57
28. Pagano, R. R., Conner, J. T., Bellville, J. W. *et al.* (1978). Lorazepam, hyoscine and atropine as i.v. surgical premedicants. *Br. J. Anaesth.*, **50**, 471
29. Pandit, S. K., Heisterkamp, D. V. and Cohen, P. J. (1976). Further studies of the anti-recall effect of lorazepam. A dose–time–effect relationship. *Anesthesiology*, **45**, 495
30. Reves, J. G., Corsson, G. and Holcomb, C. (1978). Comparison of two benzodiazepines for anaesthesia induction: midazolam and diazepam. *Can. Anaesth. Soc. J.*, **25**, 211
31. Seppälä, T., Korttila, K., Häkkinen, S. and Linnoila, M. (1976). Residual effects and skills related to driving after a single oral administration of diazepam, medazepam or lorazepam. *Br. J. Clin. Pharmacol.*, **3**, 831
32. Stoller, K. P., Belleville, J. P. and Bellwille, J. W. (1976). Visual tracking following lorazepam or pentobarbital. *Anesthesiology*, **45**, 565
33. Thornton, J. A., Dixon, R. A. and Bennett, N. R. (1978). Flunitrazepam (Rohypnol: Ro 05-4200) versus diazepam in conservative dentistry: a randomized cross-over trial. Presented at the *Vth European Congress of Anaesthesiology*, September 4–9, Paris
34. Trieger, N. (1973). Intravenous sedation. *Dent. Clin. North Am.*, **17**, 249

# 11

# Sleep and Mood:
# Measuring the Sleep Quality

P. Visser, W. F. Hofman, A. Kumar, R. Cluydts,
I. P. F. de Diana, P. Marchant, H. J. Bakker,
R. van Diest and P. A. M. Poelstra

## 11.1 THEORETICAL FRAMEWORK

A brief outline of a still incomplete psychophysiological theory, which forms the basis of our experiments, is presented in this section. This theory has two parts: one is a descriptive quantitative psychological part, the other is a physiological part based on the ideas of Karli[5] about behavioural neurophysiology.

The descriptive quantitative psychological part is based on the assumption that the effect of a disturbing influence can be measured in the morning as the change in pleasure effects of the sleep. Such measures shall have common causes with the physiological parameters, which are measures of the same 'behavioural state' of the subject. However, no strict isomorphism in the psychophysiological sense is hypothesized. It is really not meant to reduce psychological or behavioural variables to epiphenomena of the biological variables, or vice versa.

The physiological part of the theoretical framework aims at uncovering the dialogue between the organism and its familiar environment[5]. The 'motivation' of eating, drinking, sexual, temperature controlling and sleep behaviour is controlled at two different levels in the central nervous system under the guidance of the medial thalamic nuclei. The lower level is formed by the hypothalamus and the mesencephalon, and some parts of the reticular formation which aims at the control of steady states in the 'milieu intérieur'. Specific sensors measure changes in the influx/efflux relation across the cell membranes, e.g. glucose or water leading to hunger or thirst. Eating or drinking leads to restoration of the steady state. The same is true for the restoration of temperature, sleep and sexual activity. All these restorative activities of the respective steady states give pleasure.

The higher level of the control system is formed by the limbic system, which imprints personality traits and individual experience from the life history, Thus, aspects specific for one individual and cognitive or affective aspects adapted to a given situation help to make the response specific for one subject[2].

Therefore this theoretical basis leads to a system-theoretical approach with a multifactorial design and with input and output variables of both physiological and psychological nature. Thus comparisons between purely biological responses and introspective data like mood, etc. are necessary for research on sleep and dreaming. The sleep laboratories of Brussels and Amsterdam have therefore asserted the importance of objective measures of sleep quality[3], mood[9] and performance[8, 10], and their correlations with reliable measures of the sleep EEG[4, 6] and autonomic measures such as respiration[7].

## 11.2 MEASURES OF SLEEP QUALITY

Sleep quality is described in a system (Figure 11.1) with input and output relations of mood, performance and the subjects' own impressions of sleep quality after awakening. The sleep quality is measured by a psychometric questionnaire[3] and by analogue rating scales. The Wilkinson auditive vigilance test (WAVT) and the simple reaction time (RT) were used as performance measures[8, 10]. The mood was measured by five mood dimensions of the profile of mood states or POMS[9]. The disturbing influences studied by our group are both exogenous ones like traffic noise, films, environmental temperature, and endogenous ones like depression and menstrual cycle[2].

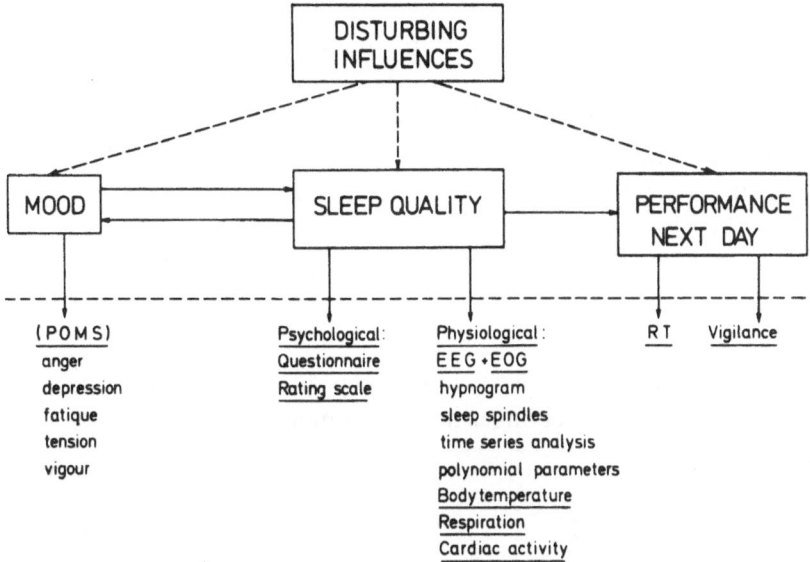

**Figure 11.1** System for measuring sleep quality. Disturbing influences can be either exogenous (noise, film, environmental temperature), or endogenous (depression, menstrual cycle)

## 11.3  EXPERIMENTAL RESULTS

### 11.3.1  Traffic noise as sleep-disturbing factor

As a first means of disturbing sleep, traffic noise was studied in a laboratory situation in two male subjects of 22 and 27 years of age. They slept for two periods of five consecutive nights in a laboratory room. These two experimental periods were preceded and separated by one week of sleep at home. The first laboratory week was meant as an adaptation period without traffic noise. During the second period in the laboratory, however, the subject was exposed to a continuous disturbance of traffic noise at a mean level of between 60 and 65 dB (A) during the whole night. The subjects rated, for the full four successive weeks, each morning their own sleep quality by means of the 11-item sleep quality scale (Figure 11.2). In the two laboratory weeks the subjects completed every morning the WAVT and the RT tasks with a variable interstimulus time. The WAVT lasted half an hour, whereas the RT task consisted of five blocks each of 48 light flash stimuli. The interstimulus time varied between 0.78 and 3.12 seconds.

137

Figure 11.2 shows that the sleep quality scores are consistently lower under the condition of traffic noise: one-sided $t$-test $p \leqslant 0.05$. The mean RT and the standard deviation of the RT (SDT) were calculated for each block of the reaction time test. The RT values with traffic noise for both subjects were higher, $p \leqslant 0.05$, whereas no differences for SDT were found. The number of signals missed in the WAVT was calculated, and increased by 60% for both the subjects after the nights they slept under environmental noise disturbance.

It could be concluded from these results that:

(a) the values obtained for the sleep quality are sensitive to the experimental changes in environmental conditions;

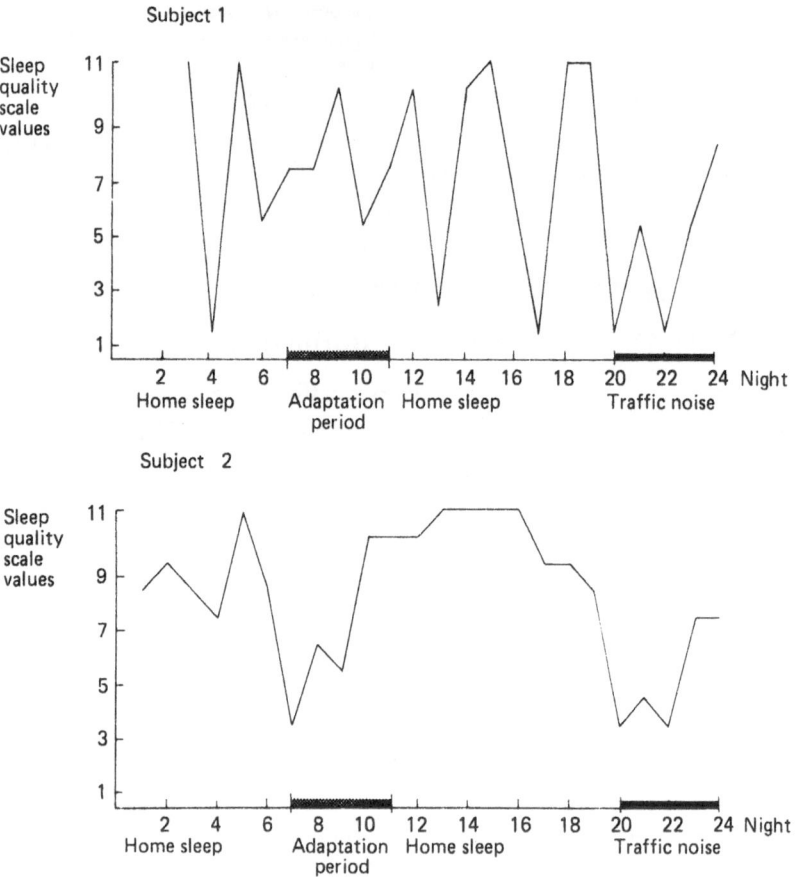

**Figure 11.2** Influence of traffic noise on sleep quality in a laboratory situation. Note: subject 1 had rather irregular sleeping habits

138

(b) the changes in sleep quality are reflected in lower performance levels, both on the mean RT and the number of missed signals in the WAVT.

### 11.3.2 Films as an external experimental disturbing influence on sleep

Films as a pre-sleep stimulation were first studied in a sample of 21 first-year male psychology students randomly selected from the population of about 250. Data of eight subjects could be used for analysis. The subjects slept for three consecutive nights in the laboratory; the first night was not analysed as an adaptation night to the laboratory situation, but the programme was quite similar to that on both the experimental second and third nights. In the adaptation night a black-and-white film was shown, on the first experimental night a travelogue colour film (neutral stimulation) and on the second experimental night a surgical colour film about a valvotomy was shown as the stressful pre-sleep stimulation.

The subjects were allowed to sleep after the end of the film, and the EEG, EMG, eye movements, respiration and heartrate were continuously recorded. The subjects were woken up 2 minutes after the beginning of the first period of paradoxical sleep (PS) or REM sleep, if possible during a bout of rapid eye movements (REM). An interview of 14 standard questions about the possible dream recollections was taken. Subjects were asked to complete three analogue scales for tension, pleasure and anxiety, which indicated the mood of the dream memories. After this they were allowed to return to sleep. Two further times they were woken up in the third PS period and in one non-REM period, but only results of the first sleep cycle shall be discussed here.

The subjects were asked about their impressions of dream and mood the next morning. Physiological parameters of the EEG and EOG signals, as sleep stages, sigma spindle density were compared between the two experimental nights with the psychological parameters. A replication experiment with five male subjects was done in which, however, they were not interviewed about the dream content. In this second series of experiments the changes in body temperature were also measured at the external meatus of the ear.

All subjects were woken up 2 minutes after the beginning of the first PS or REM period and were asked to indicate their estimates of sleep depth and sleep quality on two analogue scales and dream mood on the three analogue scales.

The EEG recordings of the first sleep cycle were classified into sleep stages by means of the automatic sleep stage analyser[6]. A third degree polynomial was fitted to the sequence of sleep stages of the first sleep cycle to get six parameters[4].

The POMS taken after the film shown and before falling asleep gave a measure for the aversiveness of the films. The subjects were allowed to sleep for the rest of the night quietly, and when awake next morning they completed again the analogue scales and the sleep quality questionnaire.

**Table 11.1   Influence of an aversive versus a neutral film on physiological variables**

|  | *Aversive* | *Neutral* | *Test* |
|---|---|---|---|
| Set of six parameters | Better sleep | Worse sleep | Hotelling $T^2$; $p \leqslant 0.05$ |
| REM latency | 62.4 min | 70.0 min | $t = 2.78$; $p \leqslant 0.05$ |
| Stage variability | 12.5 | 18.7 | $t = 2.06$; $p \leqslant 0.05$ |
| Sigma spindle density | 0.18 min$^{-1}$ | 0.36 min$^{-1}$ | $t = 1.66$; $p \leqslant 0.05$ |
| Sleep onset latency | 22.6 min | 22.6 min | n.s. |
| Sleep depth | 4.15 | 3.85 | n.s. |
| Variability of temperature (meat. ext.) | Higher | Lower | Hotelling $T^2$; $p \leqslant 0.001$ |

The profile of mood states established the 'aversiveness' of the surgical film by higher scores on both the tension and the anxiety dimensions. After the aversive film a better sleep and less tense and more pleasurable dreams were found after the first sleep cycle (Table 11.1).

The agreement percentages between the trends of the analogue scales and those of the EEG parameters after the first sleep cycle are high (see Table 11.2).

It is clear from these data that the set of six parameters, calculated from the polynomial fitted to the sleep stages of one sleep cycle, correlates with the quality and the depth of the sleep.

After waking up in the morning, four out of five subjects indicated a lower sleep quality score on the questionnaire after the aversive film night in comparison with the neutral film night. The morning result contrary to the result after the first sleep cycle is suggestive of a shift in psychological and physiological variables for sleep quality with the progress of the night.

**Table 11.2  Percentage agreement in trends of physiological and psychological data**

| *Six EEG parameters calculated from third degree polynomial* | *Analogue scales on* | |
|---|---|---|
| | *Better sleep* | *Deeper sleep* |
| 1. Sleep onset time: shorter | 71.4 | 78.6 |
| 2. Duration stages 3 and 4: longer | 50.0 | 50.0 |
| 3. Duration stages 3 and 4 until REM: shorter | 64.3 | 57.1 |
| 4. Sleep depth: deeper | 78.6 | 92.9 |
| 5. REM latency: shorter | 50.0 | 57.1 |
| 6. Sleep stage variability: less | 85.7 | 92.9 |

It is important to stress the point that both electrophysiological parameters and psychological parameters show comparable trends under the influence of exogenous stimuli. For the general practice EEG recordings are not necessary in all cases, but the indication of the patients or subjects should be made as objective as possible by means of questionnaires and/or analogue scales, which are standardized.

In another experiment with six subjects the influence of the same film as pre-sleep stimulus on mood and EEG parameters was studied. After the aversive film vigour decreased ($p \leqslant 0.05$) and fatigue increased ($p \leqslant 0.01$), the percentage of sleep stage 3 plus 4 decreased ($p \leqslant 0.03$), the total sleep time increased ($p \leqslant 0.03$) and the number of REM periods increased ($p \leqslant 0.01$).

### 11.3.3  Longitudinal study in one subject during 42 days and nights

A healthy, working female, 32 years of age, participated in this experiment for 42 days and nights, which were divided into four successive periods, P1 to P4 (Figure 11.3).

The mean sleep quality and the mean total mood disturbance (TMD) was calculated for each period. The mean sleep quality was lowest and the TMD highest during the period P3, in which the EEG, EOG and ECG were recorded with portable equipment. The mean TMD showed a striking fall, and the mean sleep quality increased after stopping the physiological recording (P4).

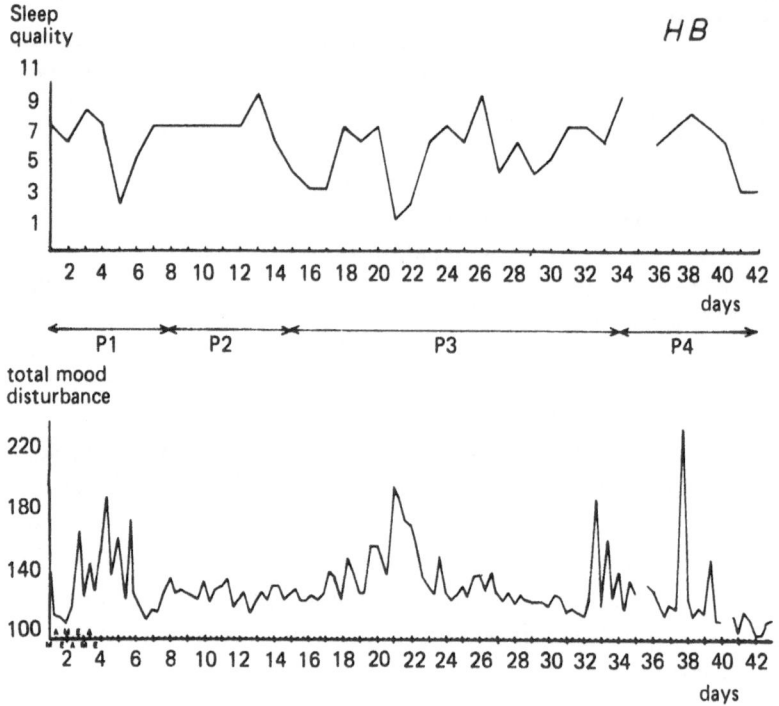

**Figure 11.3** Sleep quality and total mood disturbance of one subject during 42 days. Period P1, 7 days: sleep quality and POMS with six dimensions. Period P2, 7 days: sleep quality, POMS and half-hourly log of day's activities. Period P3, 19 days: sleep quality, POMS, day's activities log, recording of EEG, EOG and ECG with small portable recorder. Period P4, 9 days: sleep quality and POMS

Total bodily activity was calculated from the half-hourly day's activities log, which the subject filled in from day 8 until day 34 (P2 plus P3). Multiple regression analysis showed correlation of the amount of bodily activity with the sleep quality and the mood scores of the next morning only.

TMD was lower, sleep quality higher on working days than on non-working days ($p \leqslant 0.05$). Vigour was higher, fatigue and tension lower on working days. TMD was higher and sleep quality lower on a non-working day after a working day than on a working day after a non-working day ($p \leqslant 0.05$). Thus regularity of working days influenced both mood and sleep quality in a positive way.

### 11.3.4   Correlations between sleep quality and mood

In different experiments, in total more than 50 nights, a positive correlation between sleep quality and vigour and a negative correlation between sleep quality and fatigue was found after awakening. In the longitudinal study, however, negative correlations were found between sleep quality and TMD scores of the next morning: $r = -0.46$, $p \leqslant 0.05$ on non-working days, but not on working days. This could mean that the morning mood of this subject on working days was not influenced by the quality of her sleep.

### 11.4   DISCUSSION

These experimental results convincingly indicate that relations between biological measures and sleep quality can be found. This was not the case in some previous studies, which, however, did not pitch their demand high on the psychometrics of the sleep quality measures used. However, it is necessary to use other objective measures of behaviour, e.g. performance tests, beside the subject's own impressions of his or her sleep quality. Our aim of obtaining insight into both the biological and the behavioural structure of sleep quality is not yet reached, but so far these results encourage us to follow the chosen model for future detailed experiments. Though the results indicate that the biological and behavioural parameters measure the same 'state' of the subjects, the warning must be given against the separate use of either biological or psychological measures only.

### 11.5   THE SLEEP QUALITY QUESTIONNAIRES

In the experiments discussed in this paper, self-rating sleep quality questionnaires in the Dutch language were used, which were found to be strong and reliable[3].

In order to get more data and to know more about the reliability of the questionnaires replications are necessary. An English version of both the general and the specific questionnaire is given here (Tables 11.3 and 11.4). It should be noted that this English version has not been tested on strength and reliability in an English-speaking population. The positive responses are underlined.

The authors would be delighted to be informed about experiences with the questionnaires.

**Table 11.3   Sleep quality (general)**

| | |
|---|---|
| 1. I feel tired after getting up in the morning | agree / <u>disagree</u> |
| 2. I usually sleep deeply during the night | <u>agree</u> / disagree |
| 3. I often lie awake for more than half an hour before falling asleep | agree / <u>disagree</u> |
| 4. I often wake up several times during the night | agree / <u>disagree</u> |
| 5. I usually fall asleep easily | <u>agree</u> / disagree |
| 6. I usually sleep quietly | <u>agree</u> / disagree |
| 7. I think that I usually enjoy my sleep | <u>agree</u> / disagree |
| 8. I often don't sleep for more than five hours | agree / <u>disagree</u> |
| 9. I often get up during the night | agree / <u>disagree</u> |
| 10. I take a sleeping drug regularly | agree / <u>disagree</u> |

Positive responses are underlined

**Table 11.4   Sleep quality (specific)**

| | |
|---|---|
| 1. I felt energetic after getting up this morning | <u>agree</u> / disagree |
| 2. I did not sleep more than five hours | agree / <u>disagree</u> |
| 3. Last night I lay awake for more than half an hour before I fell asleep | agree / <u>disagree</u> |
| 4. I enjoyed last night's sleep very much | <u>agree</u> / disagree |
| 5. I slept deeply last night | <u>agree</u> / disagree |
| 6. I felt tired after getting up this morning | agree / <u>disagree</u> |
| 7. I woke up several times last night | agree / <u>disagree</u> |
| 8. I had difficulty in falling asleep again after waking up last night. (If you did not wake up last night answer this question with disagree) | agree / <u>disagree</u> |
| 9. I felt rested after getting up this morning | <u>agree</u> / disagree |
| 10. I fell asleep easily last night | <u>agree</u> / disagree |
| 11. I think I slept quietly last night | <u>agree</u> / disagree |

Positive responses are underlined

## Acknowledgement

We wish to thank the Medical Unit, Organon International, OSS, for lending us the portable recording unit used in the longitudinal study.

## References

1. Cluydts, R. (1974). Een intrapersoonlijk, exploratief onderzoek naar slaap-gegevens en gemoedstoestanden. Non-published licentiate thesis, Free University, Brussels
2. Cluydts, R., Deck, W., Marchant, P. and Visser, P. (1978). Chronobiology, sleep and mood. In *A.P.S.S. Proceedings of the 18th Annual Meeting*, p. 162. (Palo Alto: Stanford University)
3. De Diana, I. P. F. (1976). Two stochastic sleep quality scales for self-rating of subjects' sleep. *Sleep Rev.*, **5**, 101
4. Hofman, W. F., De Diana, I. P. F. and Kumar, A. (1977). Parameter identification of the first cycle of a human hypnogram (sleep curve) by means of a third degree polynomial. In *Sleep 1976, 3rd European Congress on Sleep Research*, p. 479. (Basel: Karger)
5. Karli, P. (1976). Les bases neurophysiologiques des processus de motivation. *J. Physiol. (Paris)*, **72**, 503
6. Kumar, A. (1977). A real-time system for pattern recognition of human sleep stages by fuzzy system analysis. *Pattern Recognition*, **9**, 43
7. Kumar, A., van Diest, R., Hofman, W. F., Visser, P., Poelstra, P. A. M. and Bakker, H. J. (1978). Sleep stage dependent parameters of respiration. Presented at the *4th European Congress on Sleep Research*, September 11–15, Tirgu Mures, Romania
8. Lisper, H. and Kjellerberg, A. (1972). Effects of 24-hour sleep deprivation on rate of decrement in a 10-minute auditory reaction time task. *J. Exp. Psychol.*, **96**, 287
9. McNair, D. M., Lorr, M. and Druppleman, L. F. *EITS Manual for the Profile of Mood States*. (San Diego: Educational and Industrial Test Services)
10. Wilkinson, R. T. (1968). Sleep deprivation: performance tests for partial and selective sleep deprivation. In L. Abt and B. Ries (eds.), *Progression in Clinical Psychology*, Vol. 8, p. 28. (New York)

# Discussion II

**Moderator: Professor J. J. Bastiaans**

**Marks:** Dr Lader, is there not an extra factor that could be brought in in the interaction between the substance and the receptor? That is to say, you would expect to get a better correlation where there was a direct interaction leading to an immediate response which terminated when the substance came off or was displaced, whereas if you had the hit-and-run type of action where there was persistent effect then you would expect to get a less good correlation. I wonder how far that might apply in the case of the benzodiazepines?

**Lader:** Yes, the best example of where you run into problems is when you are dealing with the monoamineoxidase inhibitors, where the plasma concentrations are probably quite irrelevant because once you've got a high enough concentration to knock out 85 or 90% of the monoamine-oxidase in the body, then you get the rise in the brain amines and the clinical effects which have been claimed. So I agree that that is another factor, that the interaction with the receptor must be a reversible one following the laws of mass action. However, I still feel that one can circumvent this by going for an effect, because the effect can be meaningful whether it's an irreversible or a reversible effect.

**Breimer:** I think that the work by Dr Möhler showed that in principle we are dealing, as far as the benzodiazepines are concerned, with similar types of action or similar dose–response relationships as for many other drugs. I do not think that in this case there is sufficient reason to believe that we are dealing with hit-and-run drugs.

**Oswald:** Dr Amrein felt that he could conclude that there was a correlation between the plasma levels of Rohypnol and the clinical effects. I feel that some caution should be exercised before reaching this conclusion. It would appear that his study was confounded by a number of things, particularly learning, which always is a great trouble in these tests of skilled performance. We may also have been seeing an interaction between the biological rhythm of efficiency and the drug. In the morning, after about 10 hours, it was still apparent that there was a significant differ-

146

ence between the plasma levels after the various doses, whereas there was no significant difference after about 10 hours in the performance or self-rating measures. I would suggest that that last fact is so because the subjects had learned how to do the tests and had come to be very skilful. If we then consider the subsequent evening, had the tests still been sensitive, had the subjects not learned how to do them, you might still have found a difference, because then you would be getting an interaction again in the evening between the circadian rhythm and the drug.

**Amrein:** Concerning learning, I think we have two conditions. One is the self-rating, and here there is nothing to learn – you feel as you do at the moment. The other thing was the tracing test, and there we did consider the phenomenon of learning and test whether or not people are able to learn the test. We proceeded in the following manner: you have seen that there were two tracing figures. One was always the same and the second was changed from one session to another. That means that for the first test you have the same figure and for the second never the same. We found that the result with both figures was similar and the effect of learning was minimal. Concerning the concentration at the end of 12 hours or 24 hours, I think you need a minimal concentration to have some effect, and if you are under this limit, even if you have some differences in concentration, that would not make any difference in effect.

**Saletu:** Well, I think there is another difficulty in correlation that might be a technical statistical one. We just recently finished a study with oxazepam where we did blood-level investigations as well as quantitative EEG and psychometric studies and we could only obtain a significant correlation when we averaged the Spearman rank correlations. When we tried to correlate the psychometric tests with blood levels at different times the results were completely crazy – sometimes even completely opposite. But on the first few the curves were rather similar, extremely similar as a matter of fact. And I think there are still some technical problems concerning the correlation statistics.

**Amrein:** That is absolutely correct. I think we have highest correlation between plasma concentrations over time. On the other hand, it seems absolutely clear also that the test results after 1 hour will influence the results in the next session. We saw this difficulty and to be sure that it is not a technical correlation we reduced the data for one session, that means for the maximum effect, and the correlation was always there. If you say the correlation might be only 0.5 or 0.6, I would agree with you. There are technical factors making the correlation higher than it really is. I think there is no other way possible than the way we did it, as you need the knowledge of the pharmacokinetic behaviour in the individual case.

**Nicholson:** Do you have any information on the relative sensitivity of your performance test? For instance, how long are your tests of performance impaired with 10 mg of diazepam?

**Amrein:** We pre-checked the method, and it seems that we have some response with 10 mg up to 12 hours. Twenty-four hours after, there was no positive result. It could be that the end of the clinically measurable effect of 10 mg diazepam would be some time in between. At 12 hours, some of the items were positive and others not, but after 24 hours all tests were negative.

**Nicholson:** If I've got your data right then, the persistence of impaired performance with Rohypnol (1 mg) is much less, a half, that of diazepam.

**Amrein:** Yes, I think this is due to the considerable phase of distribution of the Rohypnol, which I think is unique for a benzodiazepine.

**Breimer:** Before everybody rushes out of the auditorium to measure saliva, we have experience with that, and I think one should be very cautious. It depends very much which phase of the curve you are measuring whether you get a good correlation or not. If you measure during the elimination phase, in our experience, then it gives you quite a good correlation for a number of drugs. But generally during the absorption phase there are huge fluctuations, and we believe you cannot do without plasma levels. We first thought that these fluctuations were due to the fact that, if you give a drug orally, it is absorbed through the mucosa, but also if you give a drug rectally you get similar fluctuations.

**Lader:** Yes, I entirely agree with that. I was talking about the attempt to correlate plasma concentration or at least body concentrations with a clinical effect. Now during the absorption in the early re-distribution phases things are happening so quickly that it's almost impossible to try and get anything meaningful at that point and I would certainly agree that then you cannot use salivary concentrations. The first problem is that sometimes just after taking the drug there is some left in the mouth, especially if it is a tablet and not a capsule, and secondly the equilibrium between the saliva and the plasma is not an instantaneous one so there are complex kinetics going on. But if you are using a compound with a fairly long half-life and you wait until the early distribution phase is over then I think you have got a tool which is sometimes more convenient. We propose to use this in epidemiological studies of the 10% of the population (or whatever it is) who are taking benzodiazepines to see what sort of concentrations they have attained, and in that sort of framework it is obviously much more convenient to give them a bottle to spit into than to go round with a syringe trying to get plasma from them. But I would entirely agree there are limitations to all these techniques and we have to work within those limitations.

**Amrein:** Do you have the idea that the composition of the saliva varies according to whether salivation is provoked or spontaneous?

**Lader:** We looked at the relationship between the concentration pH and salivation rate and none of those correlations were significant. It seemed as if one could get a fairly good estimate without the complications. What you do get though is a different ratio between free plasma and saliva concentrations in different people and therefore maybe it's better as a semi-quantitative tool or as a within-individual tool rather than something from which one can derive very much between-individual information.

**Lehmann (Zurich):** I should like to corroborate Dr Visser's results that disturbed sleep or sleep under stress causes more pleasant dreams and also shortened inter-REM periods. However, I have to add that in our experience, if stress continues for a second night then dreams regain their normal unpleasant characteristic again. One should probably be careful in using this sort of assessment if only one night is being taken as a base-line night. If one continues to test the effect of drugs on dream pleasant-ness or unpleasantness, one probably needs a number of base-line nights.

**Visser:** We did not repeat our measurements in this experiment, but I think I agree with you that we should be careful about this use of stimulants on only one night. When you have this rather disturbing traffic noise every night, which is perhaps a little bit comparable, then after a few nights this influence wears off.

**Lader:** I would like to correct one misconception that some of the speakers seem to have, as is shown on their slides, and that is concerning the metabolism of temazepam. It is true that temazepam is metabolized by dealkylation to oxazepam, but that is only true for about 15% of an administered dose. The other 85% is metabolized by glucuronidation of the 3-hydroxy form, which means that in effect temazepam has no active metabolites.

**Pletscher:** I was most interested to hear that there are now objective measurements of the quality of sleep. This question has occupied us for a very long time and I want to come back to drugs and ask the panel if you see differences in the quality of sleep between the newer types of sleep-inducing agents like the benzodiazepines and the older ones like the barbiturates. Of course the pharmacological spectrum of these two classes is different. There are many differences, less enzyme induction by the benzo-diazepines, a broader spectrum of action. But my question bears on the quality of sleep. During the development of the benzodiazepines it took us a very long time before we recognized that these drugs are valuable in sleep disorders. Librium (chlordiazepoxide), the first one, was discovered just because it did not produce sleep in the animal. It took away the rage, the anxiety, you could say it made wild animals tame, but without sleep, without sedation, and then just by chance it was found in clinical practice that some of the benzodiazipines produce sleep. We know that there must certainly be differences in the mechanism of action, also, but I am wonder-ing whether you can tell us whether there are differences in the quality of sleep, as Professor Visser has been mentioning.

**Nicholson:** Well, I have never worked on barbiturates because I was in my nappies when that was being done, so I am entirely in the benzodiazepine era and I have never compared the two drugs at all.

Most certainly in the work that we were doing, we were trying to find an effective hypnotic without residual effects, and I think that in the investigation of hypnotics subjective assessments have played too big a part. One of the problems of subjective assessments is that it leads to too high doses of the drugs being used and the individuals like to wake up the following morning having had a really good night's sleep plus a really good hangover and feeling euphoric all day long. This is the development of a good hypnotic on subjective assessment. I think we now have to go for a new era of drugs and hypnotics. We don't want powerful hypnotics, it is a matter of education of the individual that they should have a hypnotic with a limited effect, such as diazepam and temazepam.

**Pletscher:** Would you call them hypnotics?

**Nicholson:** Well, they most certainly reduce awakenings in the night and reduce sleep onset latency. I personally feel that the hypnotics that we use are far too powerful, and that we should move to these drugs like diazepam and temazepam as the first approach in the sleep problem. We should not use these powerful ones, because they are being used for their residual anxiolytic effect.

**Erdmann:** Dr Korttila's excellent paper showed the different actions of flunitrazepam and diazepam. As you know for anaesthesia we are looking for an intravenous agent to give us enough analgesia and enough hypnotic action and still have the patient awake after an hour. Ketamine was proposed for this, but ketamine has a very bad psychotomimetic side-effect, with bad dreams and horror trips; combining ketamine with diazepam reduces the amount of these side-effects, but diazepam has this undesirably long action. So we are looking for a drug which we can use in order to reduce the bad side-effects of ketamine, including the centrally induced circulatory effects which are also abolished by combination with diazepam, and still have not such long action. Now you mentioned that you could reduce the dosage of flunitrazepam to very small amounts and still have the effects described. You said you are using 0.01 mg per kg, that means 10 $\mu$g per kg. If you reduced this dosage, let us say to 5 $\mu$g per kg, do you think it would still be a usable agent in combination with ketamine to suppress dreams and antagonize circulatory effects?

**Korttila:** Yes, you are right that one should have an agent which has an analgesic effect and at the same time induces sedation or even sleep. The popularity of these sedative drugs has largely resulted from the fact that they don't affect the cardiopulmonary system so much as the intravenous analgesics do. It has been shown that flunitrazepam is better than diazepam in abolishing the after-effects of ketamine, and according to Dundee lorazepam is still better in this respect. I have not so far seen reports as to whether lower doses of flunitrazepam are able to abolish or reduce the after-effects of ketamine.

**Ingvar:** Professor Visser, I was, and I think many of us were, quite astonished to see this rather paradoxical effect upon the dream experience following a horror film that you had shown to your subjects. I wondered if anything is known about the effects of pain upon sleep, pain impressions right before you go to sleep, because they may have the same effect, perhaps, so that these feelings of discomfort might be operating through an endorphin system. I also wanted to ask whether you analysed how these people experienced the film? Did they really feel that the film was horrible or didn't they perhaps enjoy it secretly?

**Visser:** We did not study pain. We studied some other influences. There are some reports in the literature about physical exercise, about psychiatric treatment in the evening which give some indication in the same direction as what we found. I don't have any information about pain, but I could make a proposal for laboratory studies, as one of my colleagues is studying pain. The difficulty is, of course, to know how long to give it so that it would not appear too sadistic. In answer to your second question, we actually measured the changes in mood on the POMS and we interviewed all our subjects about their impressions of the film. We did that next day and we had a group of about 25 students in whom we studied the effects of 15 different films, so that we would be sure to know that we chose the most aversive film that we could find. The one we chose was a film of thoracic surgery, and the subjects were psychology students, and I am pretty sure they found it aversive and not, as far as we can tell any way, attractive from a sadistic point of view.

**Hartelius:** Concerning such aversive themes, tragedies, thrills and such like, people use these, because they have paradoxical effect of course: 'Cheer up, gentlemen, things cannot get worse', which is quite stimulating. But your students must be young people with impaired sleeping capacity, and we know that young people have the capacity to abreact a psycho-cathartic effect during sleep. In elderly patients the same set-up may rather cause exhaustion and fatigue, and may also have a more disturbing influence on the night's sleep. Have you any experience of the same set-up in elderly people?

**Visser:** No, we have not. Most of our subjects were young; not all of them were students but they were mostly under 25 years of age. The fact remains that sleep was better in the first cycle, but over the whole night it was actually worse – I hope that was clear.

**Lawson (Glasgow):** Dr Korttila, you showed a good positive correlation between residual effects of benzodiazepines and the age of your patients. Do you think this was a true association or was it perhaps confounded by something like cigarette smoking? Would you care to speculate as to the mechanism of the association?

**Korttila:** I think there are different mechanisms because as regards diazepam there is an age-related residual effect after some hours, so that the explanation may be pharmacokinetic. As regards nitrazepam, the older patients tend to perform worse the morning after they have received the drug, although they have the same amount in their blood. With fluni-trazepam, there may be a small increase in the half-life of the drug with age, but I doubt whether cigarette smoking contributes. I think older patients may be more sensitive and also their levels of flunitrazepam may be a little higher.

**Priest:** In response to Professor Visser's interesting presentation, I should like to mention that at St Mary's Hospital in London we are still trying, in spite of Dr Nicholson's chastening remark about subjective reports, to develop a subjective scale that is dependable. We are not using analogue scales because some patients have difficulty with these. We are designing the questionnaire primarily for hospitalized patients and so we are using the usual order of rating scales. We now have reliability data on medical patients, surgical patients, psychiatric patients and on volunteers and we shall be publishing the results shortly.

# Section III

# Treatment of Sleep Disorders
# 2. Clinical Experience

Moderator: Professor I. Oswald

# 12

# Comparative Studies with Hypnotics

## E. Wickstrøm

### 12.1 MATERIALS AND METHODS

In order to gain an impression of the hypnotic effect of flunitrazepam, the drug was compared with other commonly used hypnotics, such as aprobarbital, nitrazepam, methaqualone-diphenhydramine, flurazepam and also placebo.

I chose to study possible hypnotic effects of the medications on patients who were to undergo an operation the following day.

The variables chosen for these studies, such as length of time before onset of sleep, duration of sleep, number of spontaneous awakenings at night and condition upon awakening, are the ones that are frequently applied to clinical studies of hypnotics.

**Table 12.1  The variables chosen for the studies**

Length of time before onset of sleep
Duration of sleep
Number of spontaneous awakenings
Patients' condition upon awakening
Patients' subjective evaluation of the quality of sleep
Investigators' appraisal of the quality of sleep
Side effect

The different data (Table 12.1) were recorded on the basis of the patients' statements.

The studies were conducted in the period 1971–5, partly at Haugesund sykehus and partly at Sentralsykehuset i Rogaland, Avdeling Stavanger sykehus, and included 250 surgical patients in a pilot study and 1063 patients in double-blind trials. Patients accustomed to regular use of high doses of hypnotics (more than twice the recommended average dose) or taking other psychotropic drugs, and those with acute active pathology or psychiatric disorder, alcoholism and drug addiction were omitted from the studies. No other hypnotics were administered for at least 2 days prior to the start of the test.

The patients were given a tablet containing the drug at 9.0 p.m. the evening before an operation and examined the following morning between 6.30 and 8.0 a.m. before getting the premedication. Their statements were compared with the nurse's observations. In one group, the hypnotic effect during the first 3 days of the postoperative period was studied.

The patients, who were in the surgical or gynaecology wards, had been told that they would participate in a study of various hypnotics. They had also been told that they were entitled to refuse such participation.

The medications used in each group were identical, but varied between the various tests at times, for instance, when comparing flunitrazepam and flurazepam in a triple-blind test.

The medications were made available by F. Hoffmann-La Roche & Co. Ltd, and the company was also responsible for randomization.

The codes were not revealed until the trial had been concluded, and the data were analysed statistically by the Section of Electronic Data Processing, University of Bergen. $\chi^2$-Test was applied.

## 12.2 RESULTS

### 12.2.1 Flunitrazepam (Rohypnol), 1 mg, 2 mg, 3 mg

A pilot study, conducted in the spring and summer of 1971 with flunitrazepam, revealed that this medication had an hypnotic effect in each of the three dosages. One hundred patients received a dose of 1 mg, 100 2 mg and 50 received 3 mg. This investigation showed that the compound had an excellent hypnotic effect at each of the dosage levels. With increasing doses the length of time before onset of sleep was shorter, the duration of sleep longer and the number of spontaneous awakenings fewer. The quality of sleep was clearly better with 2 mg than with 1 mg, but no differences in side-effects were found between these two groups. While a slight further increase in the quality of sleep was noted when 3 mg was given, this was accompanied by a definite increase in the side-effects.

### 12.2.2   Flunitrazepam versus aprobarbital (allypropymal)

Eighty patients undergoing surgery were studied in a double-blind trial. Flunitrazepam 2 mg and aprobarbital 100 mg were each administered to 40 patients. The youngest patient was 15, the eldest 90 years old.

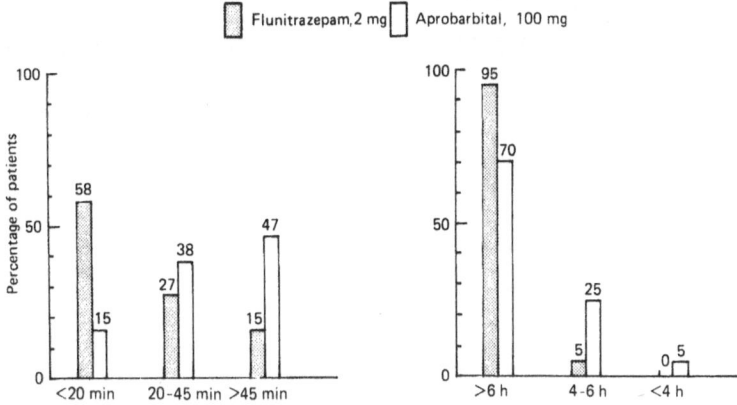

**Figure 12.1** Length of time before onset of sleep in patients receiving flunitrazepam or aprobarbital

**Figure 12.2** Comparison of duration of sleep in patients receiving flunitrazepam or aprobarbital

Time to fall asleep was shorter in the group receiving flunitrazepam ($p < 0.001$) (Figure 12.1).

The duration of sleep was longer with flunitrazepam than with aprobarbital ($p < 0.05$) (Figure 12.2).

The number of spontaneous awakenings were fewer ($p < 0.05$) (Figure 12.3).

The patients' condition upon awakening in the morning, i.e. refreshed or tired, was slightly different in the two groups ($p > 0.05$) (Figure 12.4).

The patients' subjective evaluations of the effect of the two drugs were nearly identical with the results arrived at by the investigator; both preferred flunitrazepam ($p < 0.001$ in both cases) (Figures 12.5 and 12.6).

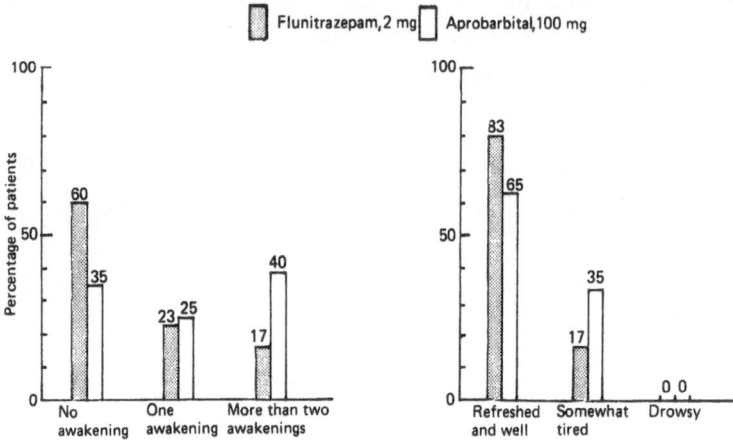

**Figure 12.3** Number of spontaneous awakenings in patients receiving flunitrazepam or aprobarbital

**Figure 12.4** Patients' condition upon awakening after receiving flunitrazepam or aprobarbital

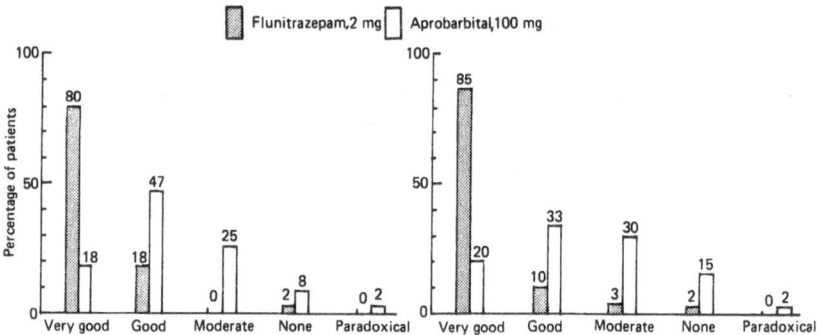

**Figure 12.5** Patients' subjective evaluations of the quality of sleep after receiving flunitrazepam or aprobarbital

**Figure 12.6** Investigators' appraisal of the quality of sleep in patients receiving flunitrazepam or aprobarbital

### 12.2.3 Flunitrazepam versus nitrazepam

Flunitrazepam 2 mg and 4 mg were compared with nitrazepam 5 mg and 10 mg. The trial included 200 patients, divided into four groups of equal size, and the patients were between 13 and 82 years of age. Group A received flunitrazepam 2 mg and one placebo, group C flunitrazepam 2 mg twice, group B nitrazepam 5 mg and one placebo and group D nitrazepam 5 mg twice, which means that all patients got two identical tablets with respect to weight and appearance, and both were taken at the same time.

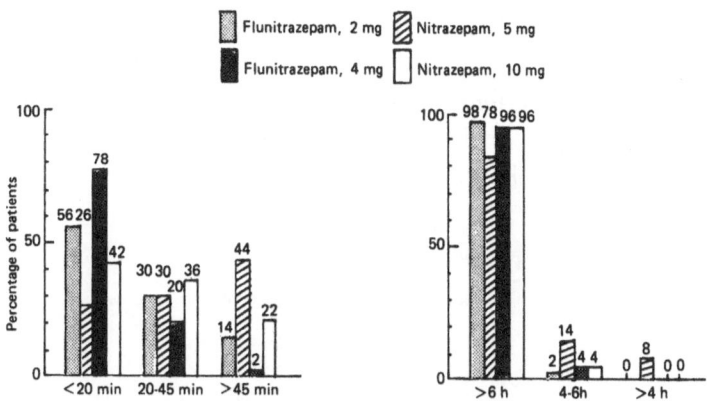

**Figure 12.7** Length of time before onset of sleep in patients receiving flunitrazepam or nitrazepam

**Figure 12.8** Comparison of duration of sleep in patients receiving flunitrazepam or nitrazepam

Figure 12.7 shows the length of time before onset of sleep. Flunitrazepam 2 mg gave a shorter sleep latency than did nitrazepam 5 mg ($p < 0.01$). Flunitrazepam 2 mg also seemed to be more active than nitrazepam 10 mg, but the difference was not statistically significant ($p > 0.05$). Flunitrazepam 4 mg gave the shortest sleep latency. Compared with nitrazepam 10 mg the difference was statistically significant ($p < 0.01$).

The registration of total sleep showed that flunitrazepam 2 mg gave a longer lasting sleep than nitrazepam 5 mg ($p < 0.01$). However, there was no difference between flunitrazepam 2 mg and nitrazepam 10 mg respectively and flunitrazepam 4 mg and nitrazepam 10 mg (Figure 12.8).

Flunitrazepam 2 mg gave fewer spontaneous awakenings than nitrazepam 5 mg ($p < 0.01$). Between flunitrazepam 2 mg and nitrazepam 10 mg there was no significant difference ($p > 0.05$). However, there was a significant difference between flunitrazepam 4 mg and nitrazepam 10 mg ($p < 0.05$) (Figure 12.9).

The condition of the patients upon awakening was not significantly different ($p > 0.05$) (Figure 12.10).

Concerning the quality of sleep, the patients as well as the investigator favour flunitrazepam. The differences were significant between flunitrazepam 2 mg and nitrazepam 5 mg ($p < 0.01$ and $p < 0.001$ respectively), flunitrazepam 2 mg and nitrazepam 10 mg ($p < 0.05$ and $p < 0.01$ respectively), and between flunitrazepam 4 mg and nitrazepam 10 mg in both evaluations ($p < 0.001$) (Figures 12.11 and 12.12).

Ten per cent of patients using flunitrazepam 2 mg complained of hangover as compared with 14% of patients using nitrazepam 5 mg; after flunitrazepam 4 mg, 22% of patients complained of hangover

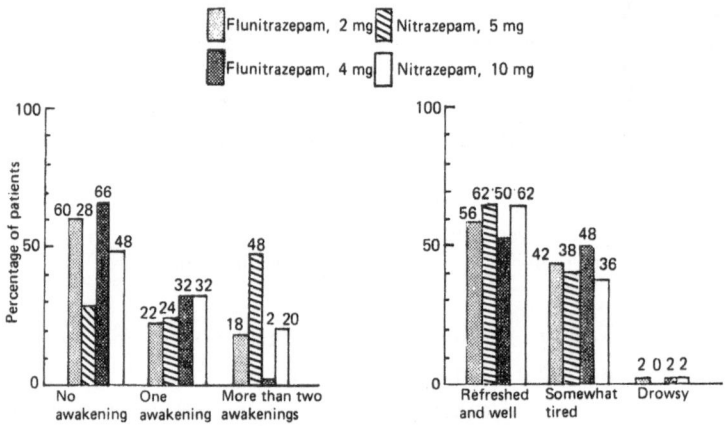

**Figure 12.9** Number of spontaneous awakenings in patients receiving flunitrazepam or nitrazepam

**Figure 12.10** Patients' condition upon awakening after receiving flunitrazepam or nitrazepam

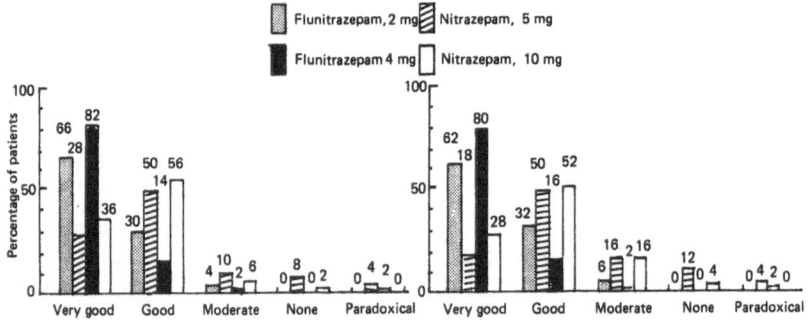

**Figure 12.11** Patients' subjective evaluation of the quality of sleep after receiving flunitrazepam or nitrazepam

**Figure 12.12** Investigators' appraisal of the quality of sleep in patients receiving flunitrazepam or nitrazepam

160

and dizziness, and 18% of the patients using nitrazepam 10 mg experienced hangover and headaches.

### 12.2.4  Flunitrazepam versus placebo

The comparison between flunitrazepam 2 mg and placebo involved 50 patients. One patient was excluded for a non-medical reason. The patients were between 16 and 81 years of age.

The length of time before onset of sleep was far shorter in the flunitrazepam group ($p < 0.001$) (Figure 12.13).

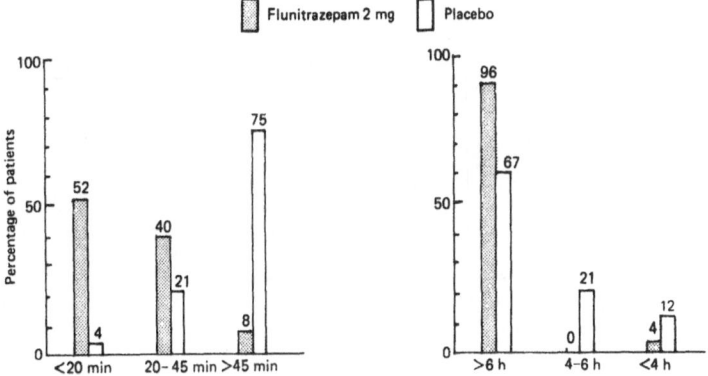

**Figure 12.13**  Length of time before onset of sleep in patients receiving flunitrazepam or placebo

**Figure 12.14**  Comparison of duration of sleep in patients receiving flunitrazepam or placebo

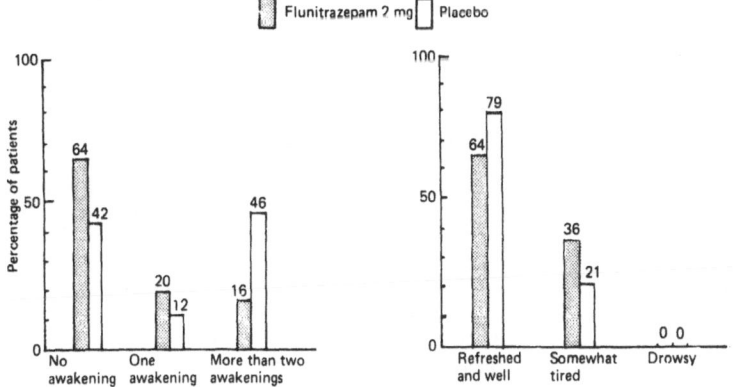

**Figure 12.15**  Number of spontaneous awakenings in patients receiving flunitrazepam or placebo

**Figure 12.16**  Patients' condition upon awakening after receiving flunitrazepam or placebo

161

The duration of sleep also showed a difference in favour of the active ingredient ($p < 0.05$) (Figure 12.14).

The reports of spontaneous awakenings are shown in Figure 12.15. Here we found no significant difference ($p > 0.05$), and we also found no statistically significant difference concerning the state after awakening – refreshed or somewhat tired ($p > 0.05$) (Figure 12.16).

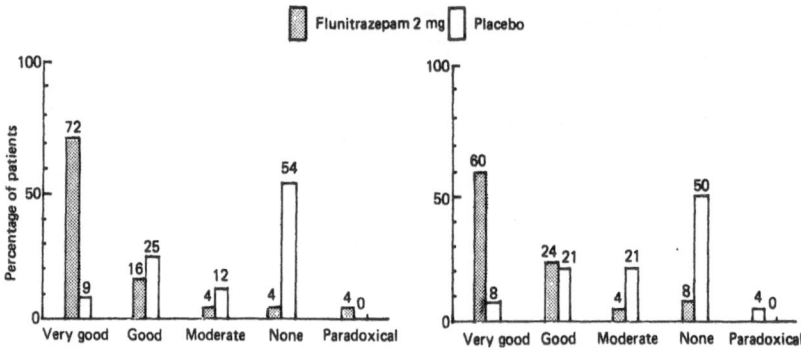

**Figure 12.17** Patients' subjective evaluation of the quality of sleep after receiving flunitrazepam or placebo

**Figure 12.18** Investigators' appraisal of the quality of sleep in patients receiving flunitrazepam or placebo

However, there was a marked difference between flunitrazepam and placebo concerning the patients' and investigators' evaluations of the sleep quality ($p < 0.001$) (Figures 12.17 and 12.18).

In the flunitrazepam group we found the incidence of side-effects to be approximately 16% against 0% in the placebo group.

## 12.2.5 Flunitrazepam versus methaqualone-diphenhydramine (Mandrax)

The trial was carried out on 400 patients. Half of the patients received a tablet containing flunitrazepam 2 mg and the other half received a tablet containing 250 mg methaqualone and 25 mg diphenhydramine.

The sleep latency was much shorter with flunitrazepam than with the combination ($p < 0.001$) (Figure 12.19).

Comparison of the duration of sleep in the two groups showed that flunitrazepam gave a longer lasting sleep than did the combination ($p < 0.001$) (Figure 12.20).

The number of spontaneous awakenings during the course of a night's sleep is shown in Figure 12.21. Flunitrazepam gave fewer awakenings than the combination ($p < 0.001$).

Concerning the condition of the patients upon awakening, the results from the two groups were nearly identical ($p > 0.05$) (Figure 12.22).

The subjective evaluations of the improvement in the sleep quality by the patients and investigators are summarized in Figures 12.23

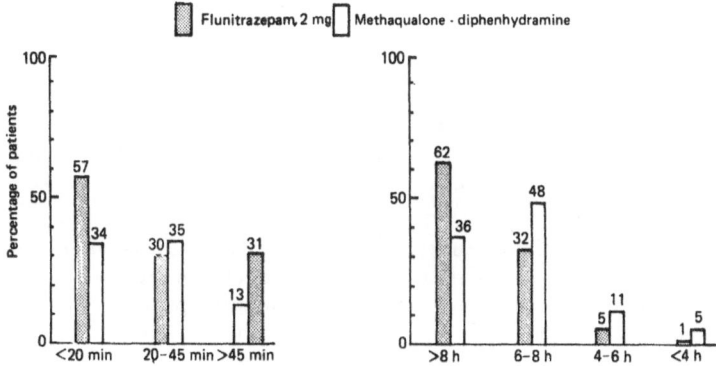

**Figure 12.19** Length of time before onset of sleep in patients receiving flunitrazepam or methaqualone-diphenhydramine

**Figure 12.20** Comparison of duration of sleep in patients receiving flunitrazepam or methaqualone-diphenhydramine

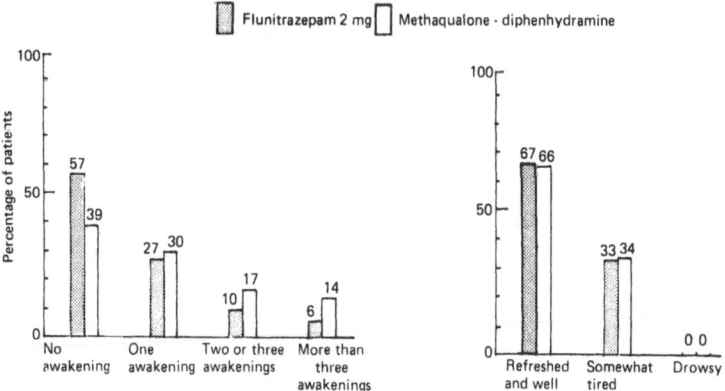

**Figure 12.21** Number of spontaneous awakenings in patients receiving flunitrazepam or methaqualone-diphenhydramine

**Figure 12.22** Patients' condition upon awakening after receiving flunitrazepam or methaqualone-diphenhydramine

163

and 12.24. According to these evaluations, flunitrazepam gave a better quality of sleep than did the combination ($p < 0.001$ and $p < 0.01$) respectively.

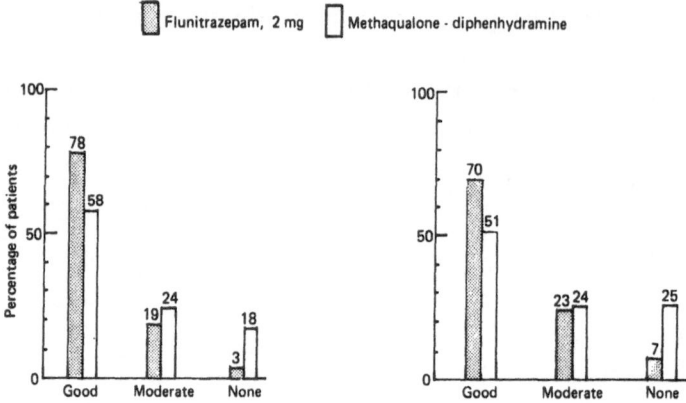

**Figure 12.23** Patients' subjective evaluations of the quality of sleep after receiving flunitrazepam or methaqualone-diphenhydramine

**Figure 12.24** Investigators' appraisal of the quality of sleep in patients receiving flunitrazepam or methaqualone-lone-diphenhydramine

Fewer side-effects were observed with flunitrazepam 2 mg and these consisted of a moderate degree of hangover and a brief but hardly troublesome dizziness upon arising.

### 12.2.6 Flunitrazepam versus flurazepam

Two hundred and thirty-three patients entered this trial which compared flunitrazepam 2 mg (114 patients) with flurazepam 30 mg (119 patients). The patients were between 16 and 81 years of age.

The length of time before onset of sleep was shorter in the group receiving flunitrazepam than in the flurazepam group ($p < 0.001$) (Figure 12.25).

The total sleeping time was longer with the former compound than with flurazepam, but no significant difference ($p > 0.05$) (Figure 12.26).

Concerning the number of spontaneous awakenings, there was no significant difference ($p > 0.05$) and also no difference ($p > 0.05$) concerning the condition upon awakening (Figures 12.27 and 12.28).

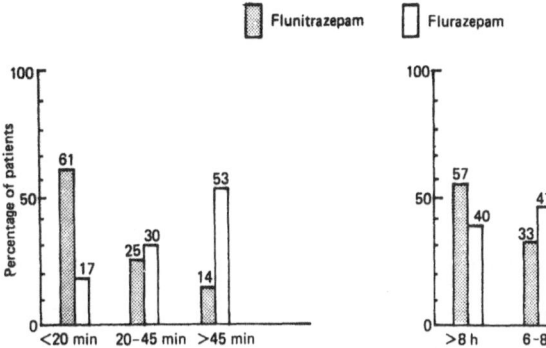

**Figure 12.25** Length of time before onset of sleep in patients receiving flunitrazepam or flurazepam

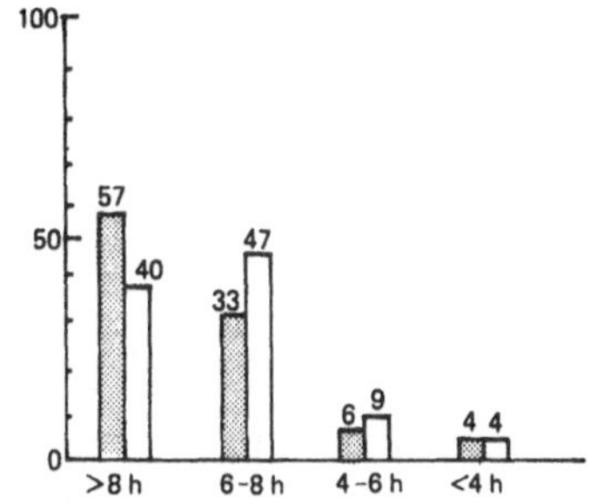

**Figure 12.26** Comparison of duration of sleep in patients receiving flunitrazepam or flurazepam

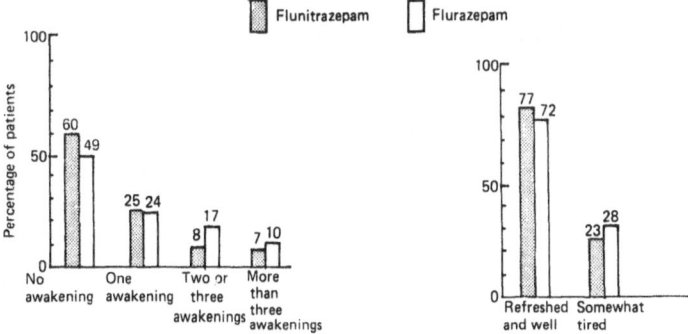

**Figure 12.27** Number of spontaneous awakenings in patients receiving flunitrazepam or flurazepam

**Figure 12.28** Patients' condition upon awakening after receiving flunitrazepam or flurazepam

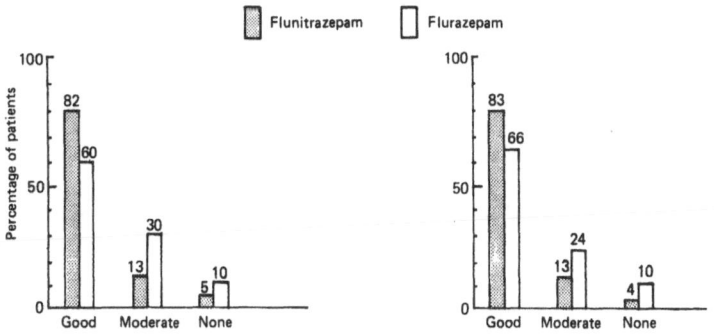

**Figure 12.29** Patients' subjective evaluation of the quality of sleep after receiving flunitrazepam or flurazepam

**Figure 12.30** Investigators' appraisal of the quality of sleep in patients receiving flunitrazepam or flurazepam

The patients and the investigators preferred flunitrazepam concerning the quality of sleep ($p < 0.05$ and $p < 0.01$ respectively) (Figure 12.29 and 12.30).

The side-effects were mild and involved a moderate hangover ($p > 0.05$).

### 12.2.7 Flunitrazepam versus aprobarbital, postoperatively

The trial included 100 surgical patients (varicose vein stripping and inguinal hernias), and each patient participated in the trial for 3 days, including the day of surgery. Six patients, three in each group, were excluded from the trial for non-medical reasons. The age of the patients was between 21 and 76 years.

Flunitrazepam 2 mg gave a shorter length of time before onset of sleep than aprobarbital 100 mg postoperatively ($p < 0.01$) (Figure 12.31).

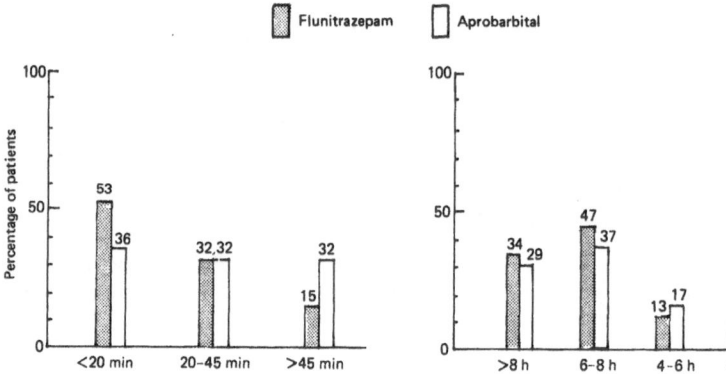

**Figure 12.31** Length of time before onset of sleep in patients receiving flunitrazepam or aprobarbital, postoperatively

**Figure 12.32** Comparison of duration of sleep in patients receiving flunitrazepam or aprobarbital, postoperatively

The duration of sleep was longer in the flunitrazepam group, the difference being significant ($p < 0.05$) (Figure 12.32).

Also the registration of spontaneous awakenings showed more favourable results in the flunitrazepam group, but the difference was not statistically significant ($p > 0.05$) (Figure 12.33).

There was also no difference ($p > 0.05$) concerning the patient's condition upon awakening in the morning (Figure 12.34).

The patients' and the investigators' evaluations of the quality of sleep are shown in Figures 12.35 and 12.36.

In the flunitrazepam group the incidence of side-effects was 1% and in the aprobarbital group 14%, mainly hangover.

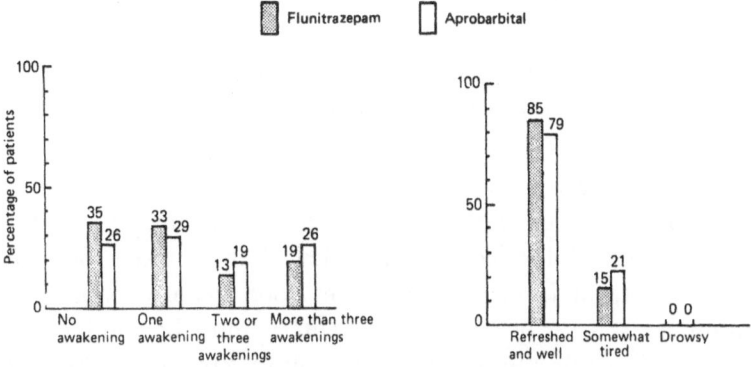

**Figure 12.33**  Number of spontaneous awakenings in patients receiving fluni-trazepam or aprobarbital, postoper-atively

**Figure 12.34**  Patients' condition upon awakening after receiving flunitrazepam or aprobarbital, postoperatively

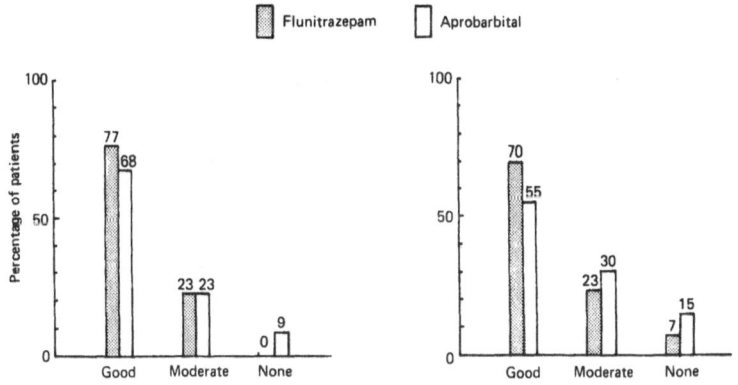

**Figure 12.35**  Patients' subjective evaluation of the quality of sleep after receiving flunitrazepam or aprobarbital, postoperatively

**Figure 12.36**  Investigators' appraisal of the quality of sleep in patients receiv-ing flunitrazepam or aprobarbital, post-operatively

Postoperatively there was some need for analgesics, all patients were given Paralgin Forte (Paracetamol 0.4 g, codeine phosphate 20 mg, promethazine hydrochloride 5 mg) as the only analgesic drug. In the flunitrazepam group 101 tablets were given during 141 days, in the aprobarbital group 102 tablets were given during the same time-span.

## 12.3 DISCUSSION

The present study showed that flunitrazepam 2 mg or 4 mg given the evening prior to an operation is a more effective hypnotic than the other products tested. A postoperative double-blind study totaling 100 patients (300 evaluations) confirmed these findings.

Clinical tests of hypnotics and their effect on sleep quality are strongly influenced by the patients' subjective feelings. In contrast to tests conducted in sleep laboratories, there are few objective criteria. What, for instance, constitutes good sleep or poor sleep? To minimize sources of possible errors, tests must be conducted as double-blind, triple-blind, or double-blind cross-over, and the number of patients involved must be substantial. In addition, a placebo group must be included.

These studies were conducted under special conditions as far as the patients were concerned. Therefore the conclusion cannot be applied immediately to other patient situations.

Nevertheless, I do feel that it is important to study the effect of the medications also under these particular conditions in order to enable their hypnotic effect to be evaluated on the broadest possible basis. The continued use of barbiturates, meprobamate or methaqualone, compounds with a small safety margin and a considerable danger of addiction, can be seriously questioned in view of the availability of newer hypnotics of at least equal effectiveness which lack these undesirable properties. The hypnotic effect of methaqualone is similar to that produced by the barbiturates and the mortality figures suggest that methaqualone intoxication, with pulmonary oedema, convulsions and vomiting, is fully as dangerous as that produced by the barbiturates.

## 12.4 CONCLUSION

The result of these studies show flunitrazepam 2 mg and 4 mg to have a good hypnotic effect; in particular, this applies to the length of time before onset of sleep and the duration of sleep. Flunitrazepam gives the fewest spontaneous awakenings, and in addition, side-effects were mild and involved only a moderate hangover and a transient dizziness upon arising.

## References

1. Wickstrøm, E. (1973). Flunitrazepam (Ro 5-4200) – et nytt hypnoticum. *Tidsskr. Nor. Lægeforen.*, **93**, 1494–1499
2. Wickstrøm, E. (1973). Clinical trials of flunitrazepam (Ro 5-4200). a new hypnotic, in comparison with aprobarbital, methaqualone-diphenhydramine, nitrazepam and placebo. Presented at the *11th Congress of the Scandinavian Society of Anaesthesiologists*, July 2–6, Reykjavik
3. Wickstrøm, E. (1974). Double-blind study of flunitrazepam (Ro 5-4200) and Mandrax. *Anaesthetist*, **23**, 90–93
4. Wickstrøm, E. (1974). Postoperativ bruk av et nytt hypnotikum – flunitrazepam. *Tidsskr. Nor. Lægeforen.*, **94**, 1053–1055
5. Wickstrøm, E. (1976). Flunitrazepam og flurazepam. En sammenlignende undersøkelse. *Tidsskr. Nor. Lægeforen.*, **96**, 1724–1726

# 13

# Pharmacological Treatment of Sleep Disorders

## C. G. Gottfries

### 13.1 INTRODUCTION

Sleep is a dynamic process and not a passive condition. From a behavioural standpoint it can be defined as a regularly recurrent state which is reversible and in which there is a greatly increased threshold for external stimulation. The neurophysiologists have specific EEG patterns when an individual is asleep, but this does not mean that the EEG pattern is the same as sleep. The reason I underline this is that when testing hypnotic drugs it is necessary not only to study neurophysiological variables but also to use questionnaires. It is of importance to rate the report of the subject and also to have trained observers' opinion of the sleep rated. Another factor to control when studying the effect of a drug upon sleep is how the person functions the next day. This capacity for performing usual tasks does not necessarily correlate to the effect of the drug on EEG sleep or the subject's own opinion.

Hypnotic drugs are extensively used and surveys have indicated that 4% of healthy people in the USA take drugs on a regular basis[4]. In hospitalized patients 80% are prescribed hypnotic drugs for treatment of insomnia[11].

When treating patients with insomnia it is of great importance to investigate whether the patient has an acceptable sleep schedule and acceptable environment for sleep. It is also of importance to

171

investigate whether there are diseases which may cause insomnia. Treatment with hypnotics is in most cases a symptomatic treatment and the specific treatment is the one which is directed to the disturbance which lies behind the insomnia.

From a clinical point of view patients with insomnia can be divided into subgroups: disintegrated sleep pattern, early morning awakening, long sleep latency, combined forms.

## 13.2 DISINTEGRATED SLEEP PATTERN

Severely disturbed sleep patterns where the whole rhythm of the night and the day is disturbed are not common. Virus encephalitis and forms of severe brain damage may cause a totally disturbed sleep pattern. As this type of disturbed sleep is chronic, pharmaceutical drugs with high risk of addiction must be avoided. Neuroleptics of the type levomepromazine are the drugs of choice. If other preparations are used, e.g. nitrazepam, these must be tried temporarily.

## 13.3 EARLY MORNING AWAKENING

The genesis of early morning awakening can be psychogenic or cryptogenic and somatogenic.

### 13.3.1 Psychogenic or cryptogenic factors

Aged people often complain of early morning awakening. The reason for this is unclear and it is too simple to say that aged people do not need as much sleep as they did when they were younger. Psychogenic or sociogenic factors can be of importance. At present we do not know if physiological age changes may cause a disturbed sleep rhythm. During recent years it has been shown that there are changes in the brain monoamine metabolism related to ageing. We also know that the monoamines have importance for sleep rhythm. Somatogenetic factors may therefore also cause early morning awakening in aged people.

In aged people one has to be careful with benzodiazepines as they may cause atonia and ataxia. It is often sufficient to treat these sleep disorders in aged people with meprobamate or chloral hydrate given at bedtime. Drugs with antihistaminic effect can also be used.

In young healthy people early morning awakening often has psychogenic causes. A middle-aged man who lives under severe stress and whose sleep routine is irregular often gets early morning awakening.

Young people with early morning awakening caused by psychogenic factors can be treated with a sedative at bedtime. Benzodiazepines are the drugs of choice. Again it must be underlined that the drug treatment is a symptomatic treatment. An improved sleep routine is, of course, the best piece of advice.

### 13.3.2  Somatogenic factors

Sleep disturbance is a significant feature of depression. Sometimes the nature of the sleep disturbance carries diagnostic and therapeutic implications. Early morning awakening is a feature of 'endogenous depression' and sleep onset difficulty is a feature of 'reactive depression'. Noyes and Kolb[13] showed that in severe depression there is no difficulty in falling asleep but the patient wakes much earlier in the morning.

Detre[2] found insomnia to be a feature in 70% of newly admitted psychiatric patients and that early morning awakening correlated with the diagnosis of depression, but sleep onset difficulty did not. In a study by Mendels and Hawkins[10], EEG was controlled during sleep in depressed patients. They found that the patients had deficiency of stage 4 and the REM stage and an excess of time awake and 'drowsy'. These disturbances were most marked in the last third of the night. In severe forms of affective psychoses also, a combined form of sleep disturbance has been reported. Early morning awakening is combined with many nocturnal awakenings. In the morning the depressed patient is often not only awake but also anxious. As a rule the sleep disturbances in affective psychoses do not respond to sedatives. More powerful drugs are required. Nitrazepam may be tried, and in patients in whom this drug has insufficient effect barbiturates or chloral hydrate may be used. In early morning waking, many patients may be helped by a sedative drug, e.g. diazepam, taken immediately after awakening.

## 13.4   LONG SLEEP LATENCY

Sleep onset difficulty is one of the most common sleep disorders and approximately 80% of sleep disorders are of this kind. The genesis varies and somatogenic and psychogenic factors, as well as the sleeping environment, may be of importance. Some subgroups can be delimited.

### 13.4.1   Late sleep onset difficulties due to pharmacological treatment

Amphetamine, caffeine and ephedrine stimulate the system of vigilance and prevent onset of sleep. Abuse of narcotics and alcohol causes serious sleep disorders which usually start with late onset of sleep. When the addicts continue with their abuse, nightmares and nocturnal awakenings will be added. The late onset of sleep caused by pharmacological treatment must, of course, if possible, be treated by withdrawal of the drug which has a stimulating effect. The treatment of addicts is a great problem and treatment of late onset of sleep is very often the same as treatment of the abstinence syndrome. In Sweden, benzodiazepines or clomethiazole in high dosage are the drugs chosen to control abstinence symptoms.

### 13.4.2   Late onset of sleep due to somatic diseases

Somatic diseases, especially those which include pain, itching, dyspnea, etc., cause difficulty in sleep onset. Pains should usually be treated with analgesics and if necessary hypnotics are added. There are many preparations which combine analgesics and hypnotics but I am not in favour of these combinations.

### 13.4.3   Late onset of sleep due to psychiatric disorders

Many psychiatric disorders, psychotic as well as neurotic, are associated with difficulties in sleep onset. Sometimes the sleep disorder also includes nocturnal awakenings. In these conditions treatment of the symptom insomnia is often the same as the specific treatment of the psychiatric disease. Treatment with sedative and hypnotic drugs is a symptomatic treatment and should therefore be

174

adapted to the specific treatment, whether this is a treatment with neuroleptics in psychotic patients or psychotherapy in neurotic patients.

### 13.4.4 Late onset of sleep due to normal psychological factors

Any individual who is under pressure due to a heavy life burden, family problems, etc., very often gets long sleep latency as the first symptom in a psychic insufficiency. The treatment of this kind of disorder is to attack the psychological and sociogenic factors which disturb the individual and hypnotic drugs should be used only temporarily. The drugs of choice are sedatives or hypnotics of the benzodiazepine type.

### 13.4.5 The importance of the sleep environment for difficulties in sleep onset

Disturbing factors in the sleep environment may cause late onset of sleep and nocturnal awakenings. Too much noise, too high room temperature, an uncomfortable bed, etc., are factors which must be investigated when treating late onset of sleep. Noise from traffic can very often be treated with earplugs.

## 13.5 COMBINED FORMS OF SLEEP DISORDERS

Combined forms of sleep disorders are seen particularly in psychotic forms of psychiatric diseases. Generally the combined forms of sleep disorders are more serious than late onset of sleep. The manic patient seems to have a reduced need for sleep. In schizophrenic psychoses both late onset of sleep and many nocturnal awakenings can be a serious problem. Psychotic conditions are frequently treated with neuroleptics. By using neuroleptics which have a central inhibiting effect, such as levomepromazine and chlorpromazine, an antipsychotic effect is obtained as well as a hypnotic effect. The benzodiazepines are here thought to be ineffective and it has been reported that schizophrenic patients have deteriorated when treated with these drugs.

## 13.6   CLINICAL VIEWPOINTS ON HYPNOTIC DRUGS

Although we have many hypnotics of differing chemical type we still have no ideal hypnotic drug. The hypnotic drugs now used produce central inhibition by which the flow of impulses to the system of vigilance is inhibited or tonus in this system is reduced. Many of the preparations used give an increased time of sleep, but EEG investigations show that this is not an increase of normal sleep. REM sleep is frequently reduced when the drug is first administered; the REM sleep time then rises to normal or slightly below normal levels and remains there until withdrawal of the drug. After withdrawal there is a sudden and sometimes marked and prolonged increase of REM sleep. Deep slow-wave sleep shows a different pattern. This is frequently unaffected by pharmacological treatment[5].

Hypnotic preparations which at present are in use in Sweden are the following: barbiturates, 'non-barbiturates', chloral hydrate, benzodiazepines, antihistamines, neuroleptics.

Barbiturates are drugs which have a rather unspecific effect. Activity both in cortical and in brain stem areas is inhibited. This has the consequence that important centres in the brain stem regulating vegetative functions are inhibited. Intoxications may therefore be fatal. Barbiturates also carry a rather high risk of addiction. To avoid the risks of barbiturates, the pharmaceutical industry has developed so-called non-barbiturate preparations. Glutethimide (Doriden), methyprylon (Noludar) and diethyldioxotetrahydropyridine (Persedon) are examples. The hypnotic effect of these preparations is somewhat weaker than that of barbiturates. The side-effects and complications of treatment with these drugs are about the same as those of treatment with barbiturates; they are therefore no step forward. At the end of the 1950s, methaqualone preparations were introduced. At first these drugs became popular, but it has been shown that intoxication with these drugs is often fatal and during recent years addictions have also been reported. It seems that the combination methaqualone–diphenhydramine is a particularly popular drug among addicts. In long-term treatment polyneuropathies have also been reported. In fact, the methaqualone preparations cannot be considered a step forward, rather a step backward.

Chloral hydrate and ethclorvynol (Placidyl) are hypnotics of non-barbiturate type with which there is long experience. These preparations are used in treatment of sleep disorders in the aged. They are as toxic as the barbiturates and the risk of abuse is similar.

The benzodiazepines have a hypnotic effect and of the preparations used in Sweden nitrazepam is the drug which has the most pronounced hypnotic effect. The benzodiazepines produce very clear-cut EEG changes, especially on long-term administration[5]. The preparations also produce some increase in sleep time, but in investigations by Hartmann[5] both the synchronized sleep time and slow-wave sleep showed a significant decrease. In fact, slow-wave sleep continued to be decreased for some time after discontinuation of medication.

The benzodiazepines are, however, a step forward from a clinical point of view. The preparations have little or no effect on centres in the brain stem and therefore intoxication carries very little risk of fatal outcome. Another advantage of the benzodiazepines is that the risk of abuse is definitely smaller than that of the barbiturates.

Antihistamines and neuroleptics have a selective inhibiting effect on the brain which can be used for hypnotic purpose. These preparations do not cause inhibition of the vegetative centre in the brain stem and therefore there is little risk of fatal outcome in intoxications. There is also no risk of addiction. These drugs have a potentiating effect on other hypnotics and they are therefore used as combination preparations. It is well known that antihistaminics have a good effect on sleep during the first one or two nights, but that the hypnotic effect thereafter disappears. If the drug is withdrawn for one or two days there is again a hypnotic effect.

Chloral hydrate (500 mg per night) was shown by Hartmann[5] to produce a decrease in sleep latency and in waking time which lasted several weeks and it also produced increased sleep time without any great change in subjective state. This may be of significance as this dose has been considered inadequate[3].

In investigations by Hartmann[5], chlorpromazine (50 mg per night) produced an increase in sleep time when first administered, but aside from this it produced almost no change in EEG sleep patterns. The lack of change with chlorpromazine was quite striking as 50 mg was the largest dose that could be taken by normal subjects on a long-term basis.

Monoamine precursors – a number of laboratories have been

investigating the role of the biogenic amines in sleep. The catechol-amines seem to be of importance for REM sleep and it appears likely that the indole amine serotonin plays a definite role in the initiation and more likely in the maintenance of sleep[1,7,9]. According to Hartmann[5], serotonin is related to the maintenance of sleep as a whole.

In animal investigations (cat) it has been shown that the 5-HT containing neurons of the rostral part of the raphe system are in-volved in sleep mechanisms. In animal experiments inhibition of the synthesis of 5-HT with *p*-chlorphenylamine leads to an insomnia which is reversed to normal sleep by injections of 5-HT-*p*. Destruc-tion of the serotoninergic neurons leads to insomnia, the intensity of which is correlated with a decrease of cerebral 5-HT. Most of the data obtained in the cat favour the hypothesis that sleep is regulated by antagonistic systems: the 5-HT neurons for inducing sleep and the catecholaminergic neurons for waking and paradoxical sleep[8].

Against the background of these biochemical findings it is of interest to consider the reports about L-tryptophan and sleep. L-tryptophan has been shown to increase the activity of brain serotonin[12].

Hartmann *et al.*[6] found in 24 insomniacs that L-tryptophan in dosage of 4 g and 5 g significantly increased sleep time and reduced awakenings and sleep latency.

In the same investigation the authors showed that 6–10 g of L-tryptophan reduced sleep latency but produced little change in sleep stages. Hartmann[5] also made a laboratory EEG study in a group of elderly women with long sleep latencies and showed a decrease of sleep latency with L-tryptophan as compared to placebo. Thus it seems that L-tryptophan may improve sleep disturbances and it can be assumed that we here have for the first time a hypnotic which improves normal sleep.

## References

1. Dement, W. C., Zarcone, V., Ferguson, J., Cohen, H., Pivik, T. and Barchas, J. (1969). Some parallel findings in schizophrenic patients and serotonin depleted cats. In D. B. Siva Sankar (ed.), *Schizophrenia – Current Concepts and Research.* pp. 775–811. (Hichville, New York: PJD Publication)
2. Detre, T. (1966). The depressive group of illnesses: sleep disorder and psychoses. *Can. Psychiatr. Assoc. J.*, Suppl. 2, 169–177

3. Goodman, L. and Gilman, A. (1970). *The Pharmacological Basis of Therapeutics*, 4th Ed. (London: Macmillan)

4. Greenblatt, D. J., Shader, R. I. and Koch-Weser, J. (1975). Psychotropic drug use in the Boston area: a report from the Boston Collaborative Drug Surveillance Program. *Arch. Gen. Psychiatry*, **32**, 518–521

5. Hartmann, E. (1974). The effects of drugs on sleep. In H. M. van Praag and H. Meinardi (eds.), *Brain and Sleep*, pp. 106–128. (Amsterdam: De Erven Bohn B.V.)

6. Hartmann, E., Chung, R. and Chien, C. (1971). L-tryptophane and sleep. *Psychopharmacologia (Berlin)*, **19**, 114–127

7. Jouvet, M. (1969). Biogenic amines and states of sleep. *Science*, **163**, 32–41

8. Jouvet, M. (1974). The regulation of the sleep–waking cycle by monoaminergic neurons in the cat. In H. M. van Praag and H. Meinardi (eds.), *Brain and Sleep*, pp. 22–38. (Amsterdam: De Erven Bohn B.V.)

9. Koella, W. P. and Czicman, J. (1963). Mechanism of the EEG-synchronizing action of serotonin. *Am. J. Physiol.*, **204**, 873–880

10. Mendels, J. and Hawkins, D. R. (1967). Sleep and depression. A controlled EEG study. *Arch. Gen. Psychiatry*, **16**, 344–354

11. Miller, R. R. and Greenblatt, D. J. (eds.) (1976). *Drug Effects in Hospitalized Patients: Experiences of the Boston Collaborative Drug Surveillance Program, 1966–1975*

12. Moir, A. T. B. and Eccleston, D. (1968). The effect of precursor loading in the cerebral metabolism of 5-hydroxyindoles. *J. Neurochem.*, **15**, 1093–1108

13. Noyes, A. P. and Kolb, L. C. (1958). *Modern Clinical Psychiatry*. (Philadelphia: Saunders)

# 14

# Role of the Sleep Laboratory in the Evaluation of Hypnotic Drugs

## C. R. Soldatos and A. Kales

### 14.1 INTRODUCTION

Pioneering studies evaluating hypnotic drugs in the sleep laboratory have been conducted by two groups: Oswald and his colleagues in Scotland and our group in the USA. In 1965, Oswald and Priest published their classical study that introduced an innovative design including three successive study conditions: placebo baseline, drug administration and placebo withdrawal[22]. The investigators were able to observe marked changes in sleep stages during drug administration and following drug withdrawal. Moreover, they were able to correlate these changes with clinical findings such as frequency and intensity of dreaming and their subjects' feelings. Specifically, they found that a rebound increase in REM sleep following drug withdrawal was correlated with an occurrence of unpleasant dreams and nightmares.

In 1969, Kales and his associates published preliminary reports[18, 20] on the evaluation of the effectiveness of hypnotic drugs which was objectively determined through sleep-laboratory recordings[7]. Another important methodological advance was their original approach in assessing not only initial and short-term drug effectiveness but whether this effectiveness was maintained over a 2-week period of continued drug administration. In later studies, the period of drug administration was extended allowing for an assessment of initial, short-term, intermediate-term and long-term effectiveness[15].

These studies established the uniqueness and importance of the sleep laboratory in the evaluation of hypnotic drugs. Their usefulness is exemplified by the fact that many pharmaceutical firms are eager to obtain sleep-laboratory evaluations of their investigational drugs. The most important and formal recognition of the usefulness of these studies is their inclusion in the Food and Drug Administration's newly published *Guidelines for the Clinical Evaluation of Hypnotic Drugs*[29]; sleep-laboratory studies are not only recommended in phases II and III, but the methodological principles derived from these studies are also applied to the general clinical trials.

In the present paper we discuss the basic characteristics of sleep-laboratory studies, their utilization in assessing the efficacy and effects on sleep stages of hypnotic drugs, and the methodological principles that originate from such studies. Results of our own studies illustrate the role of the sleep laboratory in the overall evaluation of hypnotic drugs. These studies provide a comparison between evaluations of various drugs conducted in the same laboratory using the same methodology.

## 14.2 BASIC CHARACTERISTICS OF SLEEP-LABORATORY STUDIES

There are a number of methodological strengths and weaknesses that are inherent in utilizing the sleep laboratory for the evaluation of psychotropic drugs[12,14,19]. Strengths of sleep-laboratory methodology are related to the rigorous control of experimental variables and the objective and precise measurements that are provided by this research environment. Weaknesses resulting from the use of the sleep laboratory relate primarily to the small number of subjects that can be evaluated at any one time, an inability to study special patient populations and a relative limitation in adequately assessing the side-effects of the drug studied.

In evaluating sleep, a number of potential variables that may influence the experimental conditions are not present. Thus, behaviours such as eating, drinking or physical activity are exclusively present in the waking state. In addition, there is an absence of, or at least a minimum of, external stimulation or distraction; the latter is facilitated by the sound-attenuated and temperature-controlled rooms of the sleep laboratory[12]. In addition, in the sleep laboratory investigators are able to exert a high degree of control over the various experimental variables[19]. These variables include the time for going to bed, the time for arising and therefore the total time spent in bed, as well as complete compliance for taking the experimental drug or placebo and completing the required questionnaires thoroughly and at the appropriate time. While the sleep laboratory is an artificial environment for the subject and there is an established first-night or adaptation effect[2], once the subjects have been fully adapted to the laboratory, their sleep patterns are very similar to those recorded by means of biotelemetry while they sleep at home[9].

A major advantage of sleep-laboratory studies is the fact that the data are collected by objective means and the measurements are extremely precise, on a minute-by-minute basis throughout the night[19]. This detailed information not only allows investigators to determine directions of change produced by administration of various drugs or their withdrawal but clearly provides a means of exact quantification of the degree of change.

In contrast to sleep-laboratory evaluations of hypnotic drugs, traditional clinical trials provide much less control of experimental variables and rely totally on subjective estimates, which are often unreliable, or on observations by nurse monitors, which can be quite inaccurate. However, traditional clinical trials are able to study large numbers of subjects, allow for the evaluation of special patient target groups and are an excellent means of thoroughly assessing the side-effects of given drugs, while sleep-laboratory studies are quite limited in each of these areas[19].

In sleep-laboratory studies, subjective questionnaires similar to the ones commonly used in clinical trials are completed at bedtime and every morning after awakening. The subjective data are of utmost importance since they document the presence of any side-effects and the subject's own perception of his overall sleep difficulty as well as his specific problem with sleep induction or maintenance on a nightly basis. In addition, the subjective data provide an assessment of sleep quality; for example, how restful and refreshing the sleep is every single night of the study. However, such subjective evaluations are only an adjunct to the basic information obtained through the objective sleep-laboratory measurements.

## 14.3  ASSESSMENT OF HYPNOTIC DRUG EFFICACY

The standard design that we have utilized throughout all of our protocols for evaluating hypnotic drugs is one which includes consecutive periods of placebo baseline, drug administration and placebo withdrawal. In all of these protocols, the drug and withdrawal conditions are contrasted to the placebo–baseline period. Each of these designs involves consecutive nights of study, which in turn permits uninterrupted observation and a thorough assessment; on each individual night, overall sleep difficulty (total wake time), sleep induction (sleep latency) and sleep maintenance (wake time after sleep onset) are assessed. The four basic protocols that we have utilized include two protocols for assessing the efficacy and effects of hypnotic drugs over a short-term period, one for 10 and one for 14 nights; a 22-night protocol for intermediate-term evaluation and a 47-night protocol for long-term evaluation.

### 14.3.1  Short-term sleep-laboratory studies

#### 14.3.1.1  *Standard protocols*

There are two basic protocols used: one a ten-night design with three consecutive drug nights which allow for evaluating the initial effectiveness of hypnotic drugs, and a 14-night design with seven consecutive nights for evaluating initial and short-term hypnotic drug effectiveness. The ten-consecutive-night protocol includes the first night for adaptation to the sleep laboratory, nights 2–4 for obtaining baseline measurements, nights 5–7 for evaluating the initial effects of the drug, and nights 8–10 for evaluating withdrawal effects. All ten nights are spent in the sleep laboratory. The design of this study is: PPPP DDD PPP (P = placebo and D = drug).

This ten-night protocol can be used to assess whether a drug has any hypnotic efficacy. Since most hypnotic drugs lose their effectiveness with continued use, positive findings with this protocol should be interpreted as indicating only initial effectiveness. Whether effectiveness is maintained with continued use must be tested through intermediate- or long-term studies. Negative results with the ten-night protocol suggest that, at this point, longer periods of evaluation of the drug are not indicated. However, negative results with this design do not conclusively prove that the drug is ineffective with initial use, since the sleep of insomniacs shows a considerable night-to-night variability and thus one drug night of extremely poor sleep can disproportionately bias the data.

The 14-consecutive-night protocol also includes the first night for adaptation and nights 2–4 for obtaining baseline measurements. This protocol provides for seven nights of drug administration rather than three nights for assessing the effects of the drug. Thus, on nights 5–11 both the initial and short-term effects of the drug can be assessed. On nights 12–14 the effects of withdrawing the drug are evaluated. All 14 nights are spent in the sleep laboratory. The design of this study is: PPPP DDDDDDD PPP.

The 14-night protocol was developed as an extension of our ten-night study design. It provides a maximum of information on drug effectiveness within a short period of time. Thus, it often enables the detection of an early development of tolerance, i.e. by the end of the first week of drug administration. An advantage of this study design as compared to protocols of longer duration is that its relatively short duration insures better compliance and cooperation on the part of the subjects. We feel that this design provides an excellent screening method for evaluating investigational hypnotics; it is objective, efficient, reliable and relatively inexpensive.

### 14.3.1.2 Clinical applications

Using the ten-night protocol, we found that sodium salicylamide, a common ingredient in over-the-counter sleep preparations, in doses of 650 mg and 1300 mg did not have any clearcut hypnotic effectiveness (Table 14.1). Specifically, the 650 mg dose had no effect on

**Table 14.1  Short-term hypnotic effectiveness: per cent changes of total wake time from baseline**

| Drug | Short-term drug (5–7) or (5–11) | | Drug withdrawal (8–10) or (12–14) |
|---|---|---|---|
| Sodium salicylamide, 650 mg | −0.3 | — | −2.8 |
| Sodium salicylamide, 1300 mg | −20.0 | — | −16.5 |
| Flunitrazepam, 1 mg | −5.1 | +1.6 | +60.1* |
| Flunitrazepam, 2 mg | −61.3* | −52.3† | −13.6 |
| Flunitrazepam, 2 mg | −54.0† | — | −2.0 |

$*p < 0.05$    $†p < 0.01$

efficacy parameters while the 1300 mg dose had a slight sedative effect[26]. The clinical relevance of these data is that sodium salicyl-amide is not likely to have any sedative or hypnotic efficacy when utilized in the usual clinical doses, 200–400 mg.

Using the same ten-night protocol, we found that flunitrazepam in a 2 mg dose was effective in inducing and maintaining sleep[4] (Table 14.1). We further studied the effectiveness of both the 1 mg and 2 mg doses of this drug over a 1-week period of consecutive-night drug administration, using our 14-night protocol[4]. The 2 mg dose was again found to be effective. However, the 1 mg dose was not effective; in addition, following drug withdrawal a worsening of sleep difficulty above baseline levels was noted. The clinical impli-cations of these findings are clear in terms of the 2 mg dose being clinically useful.

## 14.3.2  Intermediate-term sleep-laboratory studies

### 14.3.2.1  *Standard protocol*

The 22-consecutive-night protocol is as follows: nights 1–4, placebo laboratory; nights 5–7, drug laboratory; nights 8–15, drug home; nights 16–18, drug laboratory; nights 19–22, placebo laboratory. The first placebo night is for adaptation, and nights 2–4 are used for baseline measurements. In the sleep laboratory, nights 5–7 allow for measurement of the initial and short-term drug effects, and nights 16–18 are used to measure the intermediate-term effectiveness of the drug after 2 weeks of administration. In nights 8–15, when the subjects sleep at home, drug effectiveness and side-effects continue to be assessed through questionnaires that the subjects complete before going to sleep and upon arising in the morning. The last four nights are placebo nights and allow for measurement of any with-drawal effects.

Using this protocol, we have evaluated many hypnotic drugs to determine whether their consecutive nightly administration over a period of 2 weeks leads to a loss of their effectiveness[7,8]. This protocol also provides a confirmation of any previous findings of initial or short-term hypnotic drug effectiveness. In addition, it provides in-formation concerning sleep induction and maintenance following withdrawal of the drug after 2 weeks of consecutive nightly adminis-tration.

While the short-term protocol previously described provides information on initial and short-term hypnotic drug effectiveness, this information is of little clinical relevance when considering whether the drug has utility in the adjunctive treatment of the many patients who present with chronic insomnia. The value of the 22-night protocol is that its application provides initial information on the crucial question of whether a drug's hypnotic efficacy is maintained beyond short-term use[13, 14].

### 14.3.2.2  Clinical applications

Using our 22-consecutive-night protocol, we evaluated in an identical manner nine different hypnotic agents: chloral hydrate (Noctec), 1000 mg; ethclorvynol (Placidyl), 500 mg; flurazepam (Dalmane), 30 mg; glutethimide (Doriden), 500 mg; GP 41299 (investigational drug), 100 mg; methaqualone (Parest), 250 mg; methaqualone (Parest), 400 mg; secobarbital (Seconal), 100 mg; triazolam, 0.5 mg. These studies allowed us to compare the efficacy of these drugs over short- and intermediate-term drug administration and following drug withdrawal[8].

**Table 14.2  Intermediate-term hypnotic effectiveness: per cent changes of total wake time from baseline***

| Drug | Short-term drug (5–7) | Intermediate-term drug (16–18) | Drug withdrawal (19–22) |
|---|---|---|---|
| Chloral hydrate, 1000 mg | − 25 | − 17 | − 5 |
| Ethclorvynol, 500 mg | − 22 | − 7 | − 4 |
| Flurazepam, 30 mg | − 51‡ | − 60‡ | − 33† |
| Glutethimide, 500 mg | − 32 | + 19 | + 10 |
| GP 41299, 100 mg | − 33 | + 13 | + 20 |
| Methaqualone, 250 mg | − 48 | − 33 | − 2 |
| Methaqualone, 400 mg | − 36† | − 25 | + 6 |
| Secobarbital, 100 mg | − 55‡ | − 21 | 0 |
| Triazolam, 0.5 mg | − 41† | − 16 | + 61‡ |

*Reproduced from reference 8
†$p < 0.05$    ‡$p < 0.01$

On short-term drug administration, all nine drugs produced some degree of decrease in overall sleep difficulty, as reflected in the amount of total wake time (Table 14.2). However, this decrease was statistically significant only for secobarbital, flurazepam, meth-aqualone 400 mg and triazolam. On intermediate-term adminis-tration, there was a marked decrease in the effectiveness of most of these drugs. Only flurazepam was found to significantly improve sleep after 2 weeks of nightly administration. Following withdrawal of most of these drugs, sleep was similar to baseline levels. However, sleep continued to be significantly improved following the with-drawal of flurazepam and significantly worsened beyond baseline levels following the withdrawal of triazolam[8].

### 14.3.3 Long-term sleep-laboratory studies

#### 14.3.3.1 *Standard protocol*

The 47-consecutive-night protocol includes 28 nights of drug administration. The protocol is as follows: nights 1–4, placebo laboratory; nights 5–7, drug laboratory; nights 8–15, drug home; nights 16–18, drug laboratory; nights 19–29, drug home; nights 30–32, drug laboratory; nights 33–36, placebo laboratory; nights 37–44, placebo home; nights 45–47, placebo laboratory.

This protocol provides confirmation of any previous findings of short- and intermediate-term effectiveness of a hypnotic drug and can determine whether tolerance develops during a 1-month period of drug administration. The latter information is of clinical relevance since it is known that hypnotic drugs frequently tend to be prescribed for lengthy periods of time; in one survey, physicians estimated that in over 50% of the insomniac patients receiving hypnotic drug medication, their prescriptions extended for periods of 1 month or more[3]. These data emphasize the need for evaluating hypnotic drug efficacy for at least a 1-month period.

#### 14.3.3.2 *Clinical applications*

The 47-consecutive-night protocol has been used in three separate studies to evaluate the initial, short-term, intermediate-term and long-term effectiveness of flurazepam, 30 mg; of pentobarbital (Nembutal), 100 mg; of temazepam, 30 mg (Table 14.3)[5, 15]. All three drugs produced an improvement with initial and short-term drug administration. However, this improvement was significant only for flurazepam and pentobarbital. On intermediate- and long-term drug administration, both pentobarbital and temazepam were not effective whereas flurazepam maintained its effectiveness throughout the 4-week period of drug administration. Following withdrawal of each of the three drugs, the degree of sleep difficulty returned to baseline levels.

**Table 14.3  Long-term hypnotic effectiveness: per cent changes of total wake time from baseline**

| Drug | Short-term drug (5–7) | Intermediate-term drug (16–18) | Long-term drug (30–32) | Short-term withdrawal (33–36) | Intermediate-term withdrawal (45–47) |
|---|---|---|---|---|---|
| Flurazepam, 30 mg | − 48.6† | − 56.3† | − 42.6† | − 10.6 | − 10.0 |
| Pentobarbital, 100 mg | − 33.3† | − 1.6 | − 11.2 | − 1.6 | + 1.8 |
| Temazepam, 30 mg | − 11.7 | + 1.8 | + 2.9 | + 12.3 | + 41.0 |

*$p < 0.05$     †$p < 0.01$

### 14.3.4  Clinically relevant discoveries relating to hypnotic efficacy

Observations based on sleep-laboratory studies have resulted in three major findings relating to hypnotic efficacy[10,17]. These findings are the continued effectiveness of flurazepam that we have already discussed extensively, the carry-over effectiveness of the same drug for 2 days following its withdrawal[10] and the occurrence of an intense worsening of sleep upon withdrawal of certain benzodiazepine drugs[4,16] which we named 'rebound insomnia'[17]. These benzodiazepines are of short duration of action and in our studies were administered in single nightly doses for short-term (flunitrazepam, 1 mg; nitrazepam, 10 mg) or intermediate-term (triazolam, 0.5 mg) periods (Table 14.4). Our observations on rebound insomnia were also based on three studies conducted by other investigators; rebound insomnia occurred in Vogel's[28] and Roth's[23] intermediate-term studies of triazolam and in Adam's[1] long-term study of nitrazepam.

To explain rebound insomnia we proposed a hypothesis involving the recently discovered benzodiazepine receptors in the brain[21]. Specifically, we hypothesized that this syndrome may develop because of a delay or lag in replacement of endogenous benzo-diazepine-like molecules after the abrupt withdrawal of exogenous drugs[17].

**Table 14.4  Rebound insomnia with benzodiazepine drugs**

| Condition | Nights | Total wake time (minutes) |
|---|---|---|
| Triazolam, 0.5 mg | | |
| Baseline | 2–4 | 94.2 |
| Drug administration | | |
| Short-term | 5–7 | 51.7† |
| Intermediate-term | 16–18 | 78.2 |
| Withdrawal | 19–21 | 150.5† |
| Flunitrazepam, 1 mg | | |
| Baseline | 2–4 | 76.7 |
| Drug administration | 5–11 | 74.9 |
| Withdrawal | 12–14 | 122.8† |
| Nitrazepam, 10 mg | | |
| Baseline | 2–4 | 74.4 |
| Drug administration | 5–11 | 46.5† |
| Withdrawal | 12–14 | 103.8† |

$*p < 0.05$     $†p < 0.01$

## 14.4  EVALUATION OF EFFECTS ON SLEEP STAGES AND PHASIC EVENTS

At present, changes in sleep stages and phasic events such as the density of eye movements and sleep spindles have little known clinical relevance. However, for general scientific purposes, such data should be collected when hypnotic drugs are evaluated in the sleep laboratory. In this way, it may be possible in the future to determine clinically relevant correlates of changes in sleep stages and phasic events. One potential means of enhancing our knowledge of the importance of these physiological parameters is through the use of the computerized sleep EEG[24]. Such precise quantification in a non-invasive manner may reveal changes which are related to drug bioavailability, specific sites of action of the drug in the brain, or other pharmacological characteristics.

## 14.4.1 Short-term sleep-laboratory studies

Sodium salicylamide in both the 650 mg and 1300 mg doses did not produce any changes in sleep stages[26]. Flunitrazepam in a dose of 1 mg, which was ineffective, did not produce any suppression of REM sleep[4]. However, drug administration did result in a marked suppression of slow-wave (stages 3 and 4) sleep, which returned to baseline levels following drug withdrawal. Flunitrazepam in a 2 mg dose significantly suppressed both REM and slow-wave sleep with a return to baseline following drug withdrawal. Stage 2 sleep was significantly increased with both doses of flunitrazepam, probably in compensation for the suppression of the other sleep stages[4].

In the evaluation of flunitrazepam in a 2 mg dose, in addition to sleep stage changes, we assessed the effects of the drug on sleep-related phasic events[27]. The drug significantly suppressed body movement density and rapid eye movement (REM) density and significantly increased the density of sleep spindles. Following drug withdrawal, REM density returned to baseline levels whereas the changes in the density of body movements and sleep spindles persisted, but to a lesser degree than with drug administration[27]. These data show that the changes in sleep stages and other sleep-related physiological events do not necessarily follow the same direction and time course and are probably dose dependent. However, at least some of these changes appear to be independent of the presence of hypnotic drug efficacy.

## 14.4.2 Intermediate-term sleep-laboratory studies

Of the nine drugs we evaluated with our standard, 22-consecutive-night protocol, only glutethimide and triazolam significantly suppressed REM sleep on short-term drug administration[8]. The suppression produced by triazolam persisted on intermediate-term administration. However, only glutethimide showed a significant REM rebound following withdrawal.

Slow-wave sleep was suppressed with short- and intermediate-term administration for both flurazepam and triazolam; the only decrease which was significant was for flurazepam on intermediate-term drug administration. A significant increase in stage 2 sleep was noted on short-term administration of GP 41299 and triazolam and

191

on intermediate-term administration of flurazepam and triazolam. Following withdrawal, only the increase in stage 2 sleep with flurazepam persisted[8].

### 14.4.3  Long-term sleep-laboratory studies

Changes in sleep stages and phasic events were assessed in the 47-night studies conducted with flurazepam, pentobarbital and temazepam[5, 15]. REM sleep was significantly decreased with short-, intermediate- and long-term administration of flurazepam; it was only slightly decreased with pentobarbital and temazepam (n.s.). Following drug withdrawal, REM sleep returned to baseline levels in the flurazepam study and showed a slight rebound with temazepam and pentobarbital (n.s.).

Slow-wave sleep was significantly suppressed with flurazepam for all three drug conditions and remained suppressed but to a lesser degree following withdrawal. With pentobarbital, slow-wave sleep was significantly decreased with short-term drug administration, returned to baseline values with intermediate-term administration, was slightly above baseline values with long-term drug administration, and significantly increased following initial withdrawal. This significant increase in slow-wave sleep following withdrawal of pentobarbital is similar to our finding of a significant rebound increase in stage 3 sleep following the withdrawal of another barbiturate drug, secobarbital[11]. With temazepam, slow-wave sleep was significantly decreased with all three drug conditions; after withdrawal, slow-wave sleep returned to baseline levels.

Stage 2 sleep was increased on all three drug conditions in each of the three drug studies[5, 15]. However, the increases were not significant for pentobarbital. Following withdrawal, the increase in stage 2 sleep persisted in the flurazepam study.

Flurazepam significantly suppressed REM density throughout the 1 month of drug administration. While pentobarbital administration also decreased REM density throughout the drug period, only the changes for the intermediate-term period reached significant levels. Following withdrawal of each drug, REM density returned to baseline levels[25].

Sleep spindle density was increased throughout the period of administration of flurazepam and remained increased during the initial withdrawal period. This finding confirmed previous observations with short-term use of the same drug[6]. No change in sleep spindle density was noted either with the administration of pentobarbital or following its withdrawal[25].

Bursts of fast EEG activity were significantly increased during intermediate-term administration of flurazepam and short-term administration of pentobarbital[25]. Thus, the significant increases in bursts of fast EEG activity occurred concomitantly with the period of the peak efficacy of each drug. This finding may have some clinical importance, but no conclusions can be drawn unless more thorough studies provide additional confirmation.

## 14.5  GENERAL PRINCIPLES

From sleep-laboratory studies we have derived the following basic principles that insure the accuracy and reliability of the data obtained and its clinical application[12,14,19]:

(a)  Use a multiple, consecutive-night design to better control experimental variables and to obtain a comprehensive profile of a drug's effects across various conditions.

(b)  Allow for adaptation to the sleep laboratory and provide for an adequate baseline period.

(c)  Evaluate the efficacy of a drug with continued use, that is, over intermediate- and long-term periods.

(d)  Include a placebo–withdrawal period to detect any withdrawal effects.

(e)  Avoid drug interaction by allowing for an adequate washout period prior to a study and within a cross-over design between the evaluation of each drug.

(f)  Use insomniac subjects; normal subjects do not generally have a sufficient degree of sleep difficulty to adequately allow for evaluating the efficacy of a hypnotic drug.

(g)  Control for total bedtime, that is, do not allow *ad lib* sleep.

(h)  Analyse the data for the whole night and by thirds of the night.

Our paper has focused on the rationale and applications of certain standard protocols used in evaluating hypnotic drugs in the sleep laboratory. The use of a standard study design allows for comparisons between drugs and between data collected in different sleep laboratories. This is important since a limited number of subjects can be studied at any one time in a sleep laboratory. Thus, pooling together data collected in different laboratories is relevant and often necessary. Such pooling of data can only be accomplished if identical study protocols are used in the evaluation of a specific drug.

## References

1. Adam, K., Adamson, L., Brezinova, V., Hunter, W. and Oswald, I. (1976). Nitrazepam: lastingly effective but trouble on withdrawal. *Br. Med. J.*, **1**, 1558
2. Agnew, H. W., Webb, W. B. and Williams, R. L. (1966). The first night effect: an EEG study of sleep. *Psychophysiology*, **2**, 263
3. Bixler, E. O., Kales, J. D. and Kales, A. (1976). Hypnotic drug prescription patterns: two physicians' surveys. *Sleep Res.*, **5**, 62
4. Bixler, E. O., Kales, A., Soldatos, C. R. and Kales, J. D. (1977). Flunitrazepam, an investigational hypnotic drug: sleep laboratory evaluations. *J. Clin. Pharmacol.*, **17**, 569
5. Bixler, E. O., Kales, A., Soldatos, C. R., Scharf, M. B. and Kales, J. D. (1978). Effectiveness of temazepam with short-, intermediate- and long-term use: sleep laboratory evaluation. *J. Clin. Pharmacol.*, **18**, 110
6. Johnson, L. C., Hanson, K. and Bickford, R. G. (1976). Effects of flurazepam on sleep spindles and K-complexes. *Electroencephalogr. Clin. Neurophysiol.*, **40**, 67
7. Kales, A., Allen, C., Scharf, M. and Kales, J. D. (1970). Hypnotic drugs and their effectiveness. *Arch. Gen. Psychiatry*, **23**, 226
8. Kales, A., Bixler, E. O., Kales, J. D. and Scharf, M. B. (1977). Comparative effectiveness of nine hypnotic drugs: sleep laboratory studies. *J. Clin. Pharmacol.*, **18**, 207
9. Kales, A., Bixler, E. O. and Scharf, M. B. (1973). A comparison of home telemetry and sleep laboratory recordings with insomniac patients. *Sleep Res.*, **2**, 178
10. Kales, A., Bixler, E. O., Scharf, M. and Kales, J. D. (1976). Sleep laboratory studies of flurazepam: a model for evaluating hypnotic drugs. *Clin. Pharmacol. Ther.*, **19**, 576
11. Kales, A., Hauri, P., Bixler, E. O. and Silberfarb, P. (1976). Effectiveness of intermediate-term use of secobarbital. *Clin. Pharmacol. Ther.*, **20**, 541
12. Kales, A. and Kales, J. D. (1970). Sleep-laboratory evaluation of psychoactive drugs. *Pharmacol. Phys.*, **4**, 1
13. Kales, A. and Kales, J. D. (1975). Shortcomings in the evaluation and promotion of hypnotic drugs. *N. Engl. J. Med.*, **293**, 826
14. Kales, A., Kales, J. D., Bixler, E. O. and Scharf, M. B. (1975). Methodology of sleep-laboratory drug evaluations: further considerations. In F. Kagan, T. Harwood, K. Rickels, A. Rudzik and H. Sorer (eds.), *Hypnotics: Methods of Development and Evaluation*, pp. 109–121. (New York: Spectrum Publications)
15. Kales, A., Kales, J. D., Bixler, E. O. and Scharf, M. (1975). Effectiveness of hypnotic drugs with prolonged use: flurazepam and pentobarbital. *Clin. Pharmacol. Ther.*, **18**, 356

16. Kales, A., Kales, J. D., Bixler, E. O., Scharf, M. B. and Russek, E. (1976). Hypnotic efficacy of triazolam: sleep laboratory evaluation of intermediate-term effectiveness. *J. Clin. Pharmacol.*, **16**, 399

17. Kales, A., Scharf, M. B. and Kales, J. D. (1978). Rebound insomnia: a new clinical syndrome. *Science*, **201**, 1039

18. Kales, A., Scharf, M., Tan, T., Kales, J., Allen, C. and Malmstrøm, E. (1969). Sleep patterns with short-term drug use. *Psychophysiology*, **6**, 262

19. Kales, A., Soldatos, C. R. and Bixler, E. O. (1979). Clinical evaluation of hypnotic drugs: contributions from sleep laboratory studies. *J. Clin. Pharmacol.* (In press)

20. Kales, A., Tan, T., Scharf, M., Kales, J. and Malmstrøm, E. (1969). Effects of long- and short-term administration of flurazepam (Dalmane) in subjects with insomnia. *Psychophysiology*, **6**, 260

21. Möhler, H. and Okada, T. (1977). Benzodiazepine receptor: demonstration in the central nervous system. *Science*, **198**, 849

22. Oswald, I. and Priest, R. G. (1965). Five weeks to escape the sleeping pill habit. *Br. Med. J.*, **2**, 1093

23. Roth, T., Kramer, M. and Schwartz, J. L. (1974). Triazolam: a sleep laboratory study of a new benzodiazepine hypnotic. *Curr. Ther. Res.*, **16**, 117

24. Soldatos, C. R. (1978). Computerized sleep EEG (CSEEG) in psychiatry and psychopharmacology. Presented at the *2nd World Congress of Biological Psychiatry*, August 31 to September 6, Barcelona, Spain

25. Soldatos, C. R., Bixler, E. O., Scharf, M. B. and Kales, A. (1977). Sleep spindles, rapid eye movements and EEG fast activity and long-term administration of flurazepam and pentobarbital: further studies. *Sleep Res.*, **6**, 84

26. Soldatos, C. R., Kales, A., Bixler, E. O., Scharf, M. B. and Kales, J. D. (1978). Hypnotic effectiveness of sodium salicylamide with short-term use: sleep laboratory studies. *Pharmacology*, **16**, 193

27. Soldatos, C. R., Vela-Bueno, A., Bixler, E. O., Tan, T. L., Charney, D. S. and Kales, A. (1976). Phasic sleep events: effects of a benzodiazepine drug. *Sleep Res.*, **5**, 34

28. Vogel, G., Thurmond, A., Gibbons, P., Edwards, K., Sloan, K. B. and Sexton, K. (1975). The effect of triazolam on the sleep of insomniacs. *Psychopharmacologia (Berlin)*, **41**, 65

29. U.S. Department of Health, Education and Welfare (1977). *Guidelines for the Clinical Evaluation of Hypnotic Drugs*. (Washington, D.C.: US Government Printing Office)

# 15

## A Sleep Laboratory in a Department of Psychiatry

### W. Lehmann

#### 15.1 INTRODUCTION

The Department of Psychiatry at the University Hospital in Uppsala has a sleep laboratory which was set up in 1976. Thus we are beginners in this field; nevertheless, I shall present a brief account of how our laboratory came into existence, our activities, and our ideas on clinical and routine use of a sleep laboratory.

Our sleep laboratory is a former EEG laboratory. There are two small bedrooms, each with one bed. The separate recording room houses two Siemens Mingograph EEG machines, on which the EEG for the entire night (four channels, one EOG, two EEG, one EMG) are registered on paper, at the same time as we can record the data on a tape recorder for later processing in collaboration with the Department of Neurophysiology. The development work for computer evaluation, however, is still incomplete. Thus we evaluate our curves manually, which is a laborious procedure that takes 2–4 hours per record.

#### 15.2 USE OF OUR SLEEP LABORATORY

Up till now our sleep laboratory has been used primarily for investigation and comparison of various sedatives. Investigations carried out by us in this field include studies of the effects of dichlorazepam,

nitrazepam, oxazepam and a combination of oxazepam and promethazine. These investigations are of great practical significance,
but should ideally be alternated with theoretical studies. Projects at
present in progress include studies of the effects of vasopressin on the
memory and on REM sleep. The effects of $\gamma$-hydroxybutyric acid
are also being examined. The influence of shift work on sleep and
biorhythms is another major field awaiting attention.

Unfortunately, in the clinical use of our sleep laboratory we
encounter many practical problems of the same type as are found in
the purely pharmacological studies. It is difficult to find reliable
personnel who are prepared to work at night.

So far we have been able to use our sleep laboratory in clinical
practice only on special occasions and for specific projects. On the
other hand, our goal is a permanent organization of supervisory and
assessing personnel, which would allow sleep EEGs to be used as a
matter of routine, not solely for examination of our own psychiatric
patients, but also accepting individuals remitted from all over the
hospital.

## 15.3  A CASE REPORT

As an example of clinical use of a sleep laboratory I would quote an
edifying case, in which the sleep EEG finally led to correct diagnosis
and treatment. The patient was a 48-year-old male, remitted from
a department of internal medicine, who had heard about our interest
in sleep. The case history included a *commotio cerebri* in 1964,
colitis in 1969, diabetes mellitus discovered in 1970, and examination for arthritis, prostatitis and polyglobuli, which, however, gave
no positive results. The current complaint, since 1972, was gradually
increasing fatigue and need for sleep. The patient needed 14 hours
sleep by night and 2 or 3 by day. The fatigue was not of rapid onset,
but gradually intensified over 2 hours in the morning and evening.
He was admitted to his local hospital and examined several times.
Suspicion of hypothalamic tumour and Pickwickian syndrome
could not be verified. The investigations included encephalography.
The patient's diabetes was well controlled by chlorpropamide 0.25 g
once daily. He was wholly free of mental symptoms. The local
hospital diagnosed narcolepsy, and prescribed ephedrine, caffeine,
Tofranil and finally amphetamine, all without success. When he

came to us he had been taking amphetamine for over a year, at last in a dosage of 40–50 mg daily; nevertheless, he could not cope without his afternoon nap.

The diagnosis narcolepsy is prompted by incidence of suddenly occurring attacks of drowsiness, which may happen at any time of the day, and from which the patient awakes rested and refreshed. In addition, one or more of the following may be present: cataplexy, sleep paralysis, and hypnagogic or hypnapompic hallucinations. Our patient presented none of these.

We were finally able to reach a definitive diagnosis with the help of the sleep EEG, which in this case proved to be completely normal. In narcolepsy one would expect sleep onset REM, or at least a substantially reduced REM latency as expression of a defect in the control of REM sleep, which presumably constitutes the narcolepsy syndrome.

The solution of the mystery finally appeared in the medication with chlorpropamide (Diabines) which in this patient apparently caused this enormous need for sleep. He had been treated with Diabines since 1972, the year of onset of his drowsiness. This connection was never observed, despite all the investigations. The omission of Diabines resulted in a marked improvement with greatly reduced need for sleep. The patient was soon up from 6.0 a.m. to 11.0 p.m., with no tiredness whatever. Hypersomnia as a side-effect of chlorpropamide has, so far as we know, never been observed previously, and was reported by us to the Side-effects Committee of the National Social Welfare Board.

## 15.4 CLINICAL USE OF A SLEEP LABORATORY

In principle, at least four areas of use for a sleep laboratory are thinkable:

(a) Examination of sleep disturbances.
(b) Diagnosis of mental illness.
(c) Prognosis of mental illness.
(d) Selective sleep deprivation.

I shall now quote some examples of occasions when we believe a sleep EEG can make a valuable contribution to clinical practice.

### 15.4.1 Sleep disturbances

There are many ways of classifying sleep disturbances. Mention has been made of acute, chronic, psychogenic, biological, neurological, situational, etc. We prefer to use the following:

(a) Difficulty in falling asleep.
(b) Disruption of sleep continuity.
(c) Unduly early awakening.
(d) Disturbances in sleep physiology, i.e. changes in the physiological distribution between the various sleep stages, e.g. diminished SWS, reduced REM sleep or REM rebound. As is well known, the last may take the form of nightmares and vegetative effects. I am here thinking of increased production of hydrochloric acid in ulcer patients and increased incidence of angina pectoris in infarct patients during REM sleep. Finally, the effect of various psychoactive drugs on these conditions is also an important subject for study.

### 15.4.2 Diagnosis of mental diseases

In primary depression we almost invariably find the following four changes in the sleep EEG: short REM latency, reduced SWS, increased REM activity, disruption of sleep continuity. These parameters allow the differentiation between primary and secondary depression. Short REM latency and reduced SWS always signify the presence of an affective component in the disease. Thus, with the exception of narcolepsy, we find them only in primary depression, mania and schizo-affective disease.

### 15.4.3 Prognosis

Changes in the above-mentioned parameters also have a certain prognostic value. If, for instance, the disruption of sleep continuity in depressive patients tends markedly to reduce the total sleep time ($> 50\%$ reduction) there is a strong possibility of development of a psychotic depression.

Kupfer[4] writes in 1977: 'The rate of change of REM latency lengthening in the early stages of treatment with tricyclics appears to relate clearly to an eventual treatment response.'

Furthermore, Vogel *et al.*[7] say, also in 1977: 'Increase in late

REM% after REM deprivation is proportional to likelihood of successful response to REM sleep deprivation treatment.'

Still another example is the observation of Gross et al.[2] of decreased SWS during acute alcohol withdrawal. They suggested that the return of SWS to normal may be taken as an indication of physiological recovery.

### 15.4.4 Selective sleep deprivation

REM sleep deprivation has an antidepressant effect, as demonstrated by Vogel et al.[7] in 1975. The procedure is to rouse the patient each time he enters an REM period, which is only possible in a sleep laboratory.

Vogel showed that REM sleep deprivation significantly alleviated symptoms in individuals suffering from endogenous depression, but not in reactively depressed patients. He concluded that the efficacy of REM sleep deprivation resembled that of imipramine.

This is very interesting in theory. Cholinergic mechanisms are probably involved in both the timing of the REM periods and the physiology of depression[3]. The enhancement of cholinergic activity by infusion of physostigmine has been shown to shorten the REM latency[5]. A brief REM latency means a high pressure for REM.

On the other hand, following REM deprivation a significant fall in acetylcholine in the telencephalon of rats was reported in two studies[1,6]. The beneficial effect of REM deprivation may result from this reduction in cholinergic tonus.

So the sleep laboratory may play a major role in the treatment plan for some depressive patients, although, on the other hand, successful treatment of depression does not appear to depend solely on deprivation of REM sleep.

### 15.5 SUMMARY

Using comparatively modest resources we have set up a sleep laboratory. We believe that a sleep laboratory has its rightful place at a department of psychiatry; a multitude of tasks lies in wait. Apart from pharmacological studies there is a need for basic research on the physiology, biochemistry and function of sleep. From the point of view of psychology, dream research is associated with various interesting physiological variables. Also in the clinical sphere there are many areas where valuable contributions can be offered. Investigations of sleep disturbances, diagnosis and prognosis of mental diseases, therapy with REM deprivation are a few examples. The organization of a sleep laboratory in continual use calls for well-trained, reliable personnel, who are employed on a permanent basis. This requires substantial financial resources. Our goal is a permanent staff, and the introduction of sleep EEG as a routine method, not only for the patients of the Department of Psychiatry, but also as a service for the entire hospital.

## References

1. Bowers, M. B. *et al.* (1966). Sleep deprivation and brain acetylcholine. *Science*, **153**, 1416
2. Gross, M. M. *et al.* (1973). Experimental study of sleep in chronic alcoholics before, during and after four days' heavy drinking. *Ann. NY Acad. Sci.*, **215**, 254
3. Janowsky, D. S. *et al.* (1972). A cholinergic–adrenergic hypothesis of mania and depression. *Lancet*, **2**, 632
4. Kupfer, D. J. (1977). EEG sleep correlates of depression in man. In E. Hanin and D. Usdin (eds.), *Animal Models in Psychiatry and Neurology*, pp. 181–188. (Oxford: Pergamon Press)
5. Sitaram, N. *et al.* (1976). REM sleep induction by physostigmine infusion during sleep in normal volunteers. *Science*, **191**, 1281
6. Tsuchiya, K. *et al.* (1969). Sleep deprivation: changes of monoamines and acetylcholine in rat brain. *Life Sci.*, **8**, 867
7. Vogel, G. W. *et al.* (1975). REM sleep reduction effects on depression syndromes. *Arch. Gen. Psychiatry*, **32**, 765

# Discussion III

**Moderator: Professor I. Oswald**

**Lader:** We carried out a survey recently in 1000 people and found that something like 6% of them were taking a benzodiazepine or other sedative drug chronically and a substantial proportion of these people claimed, and it is only their claim, that they were started on these drugs when they were in hospital for an acute emergency or for surgical operation. Is the additional clinical benefit of the night's sleep on the hypnotic worth the long-term risk of starting people on many years of taking benzodiazepines and other drugs? A point that I would like to make to Dr Gottfries is that the one cause of disturbed sleep which he left out, and which in my experience of dealing with these patients is the commonest, is the continued use of the sedative compounds so that the disturbed sleep is due to a withdrawal from those compounds if the patient leaves out a dose.

**Gottfries:** There are also American investigations which show that 80% of the people who are hospitalized get hypnotic drugs. I think it is very common in general hospitals, not only when an operation is to be performed, to give patients hypnotics even when they have no sleep disturbances. So I think you are pointing to a very important thing here, and I very much agree that misuse of these drugs, of course, also in the long run causes sleep disturbances.

**Wickstrøm:** You see, when patients come to the surgery department for operation the next day, they are a little scared and they always ask for a hypnotic for the evening prior to an operation, and I think it is good to give them something. It also helps the anaesthesia, so it is good for us who are going to operate on the patient.

**Oswald:** We could ask for hands up for those who would like a sleeping pill before they have their hernia operated. I would!

**Gottfries:** May I make another comment on that. I can accept that you give this kind of patient sedative drugs or hypnotic drugs when they come to the hospital, but I think what must be stressed is that the doctor who gives the patients sedative drugs should also withdraw the drugs. It is the continuous treatment which is dangerous. You can very well give the patients sedatives for two or three nights, but then you should withdraw the treatment and I think that some doctors in hospital are not as careful as they should be.

**Wickstrøm:** Sometimes I also think we give the drugs because we have very few nurses in the hospital, and when we give the patient a drug it is easier for them during the night.

**Pletscher:** You have experience with both barbiturates and benzodiazepines. Did you see any development of tolerance with benzodiazepines? How does this compare with the barbiturates?

**Wickstrøm:** I have not looked for this.

**Pletscher:** And then you said you have a remarkable preference of your patients towards the benzodiazepines. Why? Again my question of yesterday, is the quality of sleep somehow different? Is there a euphoric component in it, or what is it?

**Wickstrøm:** The only thing I can say is that in the pilot study I gave flunitrazepam and when I asked the patients in the morning, a lot of them told me what a wonderful drug it was. 'I feel very well in the morning', they told me. There was much less hangover than using barbiturates or other commonly used benzodiazepines such as Mogadon. We have not done any vigilance tests. I think in the States they have done some, also during the evening after, and then it was seen that there was a real hangover, also during the day. But the patient did not know anything about it. He felt very well.

**Pletscher:** For how long, for how many days have you given benzodiazepines?

**Wickstrøm:** We have only given them one night when the patients are going to be operated the day after.

**Pletscher:** You have no experience with chronic treatment?

**Wickstrøm:** Yes. We have done one pharmacokinetic study for 4 weeks. Dr Amrein, perhaps you can tell us a little more about it. But the patients like the drug very much and we have also given it intravenously as anaesthesia in varicose veins which are going to be operated on both sides, and the patients like it, perhaps a little too much. They did not remember anything and perhaps it was something like euphoria or something like that. I do not know.

**Hartelius:** We made an investigation of a long-term study of flunitrazepam, together with nitrazepam and placebo, with special regard to withdrawal effects. And now a question to Dr Wickstrøm about the dosage. These were people who had been accustomed to different types of hypnotics for many months and then they were allowed during the test period to raise the dose of flunitrazepam from 1 mg, of nitrazepam from 5 mg and

204

placebo from one tablet. Of the flunitrazepam patients 15 doubled to 2 mg, and of the nitrazepam group 26 doubled to 10 mg and 27 patients raised the dose of placebo to two. I have the experience that 1 mg flunitrazepam will be enough for most of the patients. We discussed this in the corridor yesterday and I would like to hear your comment about the dosage.

**Wickstrøm:** I think in hospital it is good to use 2 mg, but at home 1 mg is enough, and if patients are using the drug when they are going on a journey or something, I think perhaps 0.5 mg would be enough if they do not have real sleep disturbances. In Belgium I heard they have used 4 mg. I am not sure if they really needed it, or if they buy 4 mg and take 1 mg or 2 mg.

**Robak (Norway):** Let me first say a few words about aprobarbital just to make it clear for you what substance it is. It is a barbiturate which can be very well compared with pentobarbital in strength and duration of action. Dr Wickstrøm, one of your comparisons was between nitrazepam and flunitrazepam and I think that your work showed very neatly that 1 mg of flunitrazepam compared very well with 5 mg of nitrazepam. But then, in the rest of your comparisons, when you compared, for instance, flunitrazepam with barbiturates, then you compared 2 mg of flunitrazepam with 100 mg of aprobarbital. But that is not a fair comparison as regards dosage I think, because comparisons have been made a long time ago between, for instance, 5 mg of nitrazepam and 100 mg of aprobarbital showing that those two substances are equal at this dosage level. So when you compare 2 mg of flunitrazepam with 100 mg of aprobarbital, then you have the double strength of flunitrazepam in this comparison. So I think that some of your results, the very favourable results as regards flunitrazepam as compared with barbiturates, are caused by your having doubled the dose from what it should have been.

**Wickstrøm:** In 1971, when I started the study, most people were using 2 mg in different studies, and if you can say that 1 mg is about the same as 5 mg nitrazepam and you are not succeeding with 5 mg nitrazepam, I think it is correct to use 2 mg. Perhaps I agree with you about the 100 mg with aprobarbital. Perhaps I should use 200 mg.

**Björqvist:** I would like to stress the point that a doctor who starts a hypnotic should also stop it. It is his special concern to stop the medication in time. I work a lot with addicts, both alcoholics and people who use sedatives and hypnotics, and they are mostly started by a doctor who wrote a prescription for 100 tablets and then he continued to give 100 tablets for several months and never thought about it further. The easiest way of avoiding this problem is never to write a prescription for 100 tablets. Fifty tablets are too many and 30 may be too many. You should start with 20, and if the patient comes back and asks for more, then you should make a very thorough investigation of what lies behind this problem, as Professor Gottfries outlined this morning.

**Ingvar:** I would like to bring up the question of age, especially the elderly patient, and the difference between barbiturates and benzodiazepines. I think it is important to realize that these two groups of drugs have very different effects upon the brain stem. The barbiturates do have an effect on vasomotor centres and the blood pressure is lowered and this I think can be deleterious in elderly patients, because of the harmful effects upon the brain circulation. That's why I think that the elderly patient requires special consideration when it comes to the choice of a hypnotic.

**Gottfries:** Yes, I agree and I think that elderly people are just those who get the hypnotic drugs and therefore much more consideration must be given to these problems. And as I said we now know that many systems in the brain have a reduced reserve capacity in the elderly and all drugs which influence these systems of course produce a different effect in elderly patients. It is often said that the receptors become more sensitive as you get older, but there is no proof for that. Investigations have shown on the contrary that the number of receptors, acetylcholine and serotonin receptors, is reduced in the aged brain. However, these age changes are not general, they are uneven. You can have one system which is sensitive to age and where you have a reduced capacity and another system which is not vulnerable to age. And this gives rise to an imbalance in these systems, and here perhaps the effect of pharmaceutical drugs may cause paradoxical effects. Take, for instance, barbiturates; sometimes in elderly people you can get aggressiveness instead of sleepiness, and that is perhaps because you influence balance in an unusual way with these drugs. So I think it is very important to study the effect of drugs on the brain of the aged.

**Ingvar:** Do the newer benzodiazepines have as little effect upon the systemic circulation as the first ones that came into general use?

**Amrein:** The new benzodiazepines on the market are at least as good concerning these vital parameters as the old ones and I hope that the newest will be even better, but there we are just at the clinical research stage. If we inject Rohypnol, for instance, there is no decrease at all in blood pressure and no influence on respiration in a dosage of 1 mg or lower. There is some effect from 2 mg given over a few minutes. That means that such effects are also dependent on the plasma concentrations. But if you use the same substance orally, you will never have as high a plasma concentration as when you give the drug intravenously, especially in a very rapid injection.

**Laihinen (Finland):** What do you think about the use of small doses of neuroleptic drugs for inducing sleep in patients in whom there is a risk of addiction? And which neuroleptics would you recommend in this respect? What do you think about levomepromazine?

**Gottfries:** Well, I think the most usual way to handle such patients is to use antihistamines or neuroleptics in small doses. Levomepromazine has a hangover effect and gives rise to hypotension, but you can try it in very low doses of course. Levomepromazine, thioridazine and chlorpromazine are the drugs which are most often used in Sweden I think.

**Breimer:** I think that some confusion has arisen with respect to the optimal dose of flunitrazepam. Yesterday during my lecture I referred to an investigation in the USA on flunitrazepam, where the Kales group showed in one study that 0.25 mg was effective and 1 mg was effective, but in a subsequent study they could not show that 1 mg was effective and only 2 mg was effective and they suggest that might possibly be due to the fact that a different pharmaceutical formulation was used. I should like to stress again that what you really want for rapid sleep induction is a pharmaceutical preparation from which the drug is very rapidly released, and if it is very slowly released you cannot expect a rapid onset of action. Apparently, the effectiveness of flunitrazepam at least partly depends on the pharmaceutical preparation used. And I think this should be investigated more carefully, because if you use a preparation from which it is rapidly absorbed, you can probably give a lower dose and I think that would be a great advantage.

**Billiard:** The most important thing to my mind is that when you are dealing with the pharmacological treatment of chronic insomniacs, you may have to give a drug for several months, even when you are well aware of the fact that you must stop as soon as possible. Therefore you must be very careful in the choice of the drug. You must be well aware, I think, that hypnotic drugs just provide a symptomatic treatment of insomnia and you need both accurate psychological evaluation and, if possible, polygraphic recordings. Accurate psychological evaluation because a subject who is just anxious benefits from tranquillizers during the daytime and does not really need hypnotics at bedtime. And on the other hand subjects with chronic ruminative depression, which Dr Kales has shown to be the most common personality pattern among insomniacs, benefit from antidepressants such as amitriptyline or doxepine and not from hypnotics. In a population of 143 subjects that we recorded polygraphically in the last years, we found that 60% of the subjects complained of long sleep latency on awakening in the morning, whereas only 30% actually showed a long sleep latency. Thus we always have to be very critical when we encounter a complaint of long sleep latency. I should also like to add that when polygraphic recordings are possible they may show special conditions such as sleep apneas, 'restless legs' syndrome and so on, which in some cases may also contraindicate hypnotics.

**Oswald:** Concerning the genesis of long sleep latency, we should always consider the fundamental question of why we fall asleep when we do. Now, we fall asleep when we do chiefly because of our biological clock and if we all flew today to India then we would have difficulty in falling asleep at 11 o'clock tonight. All too often the complaint of difficulty in falling asleep at night is caused by hypnotic drugs that are still there, making the person drowsy and lethargic in the morning, and the patient complains that she is too tired to get up in the morning because she could not get to sleep at night. The truth is that she gets up at 8 o'clock or 9 o'clock in the morning and because of that she cannot fall asleep until about 1 o'clock in the morning. If she were to get up at 6 o'clock in the morning and milk the cows, then she would have no difficulty in falling asleep at 10 o'clock at night. Not long ago a boy of 16 was referred to me who had been receiving hypnotic drugs since the age of eight. He was in a residential institution for maladjusted boys and the complaint was that he could not fall asleep until 2 o'clock in the morning. He did not have to start his lessons in this special school until 10 o'clock in the morning and he got up at about 9.30 a.m. So, of course, he did not fall asleep until 2 o'clock in the morning. I had him under my care and we made him get up at 6.30 every morning and he had no trouble after that in falling asleep without any drugs at all at 10 o'clock at night. So I think we must remember the importance of the biological rhythm, because it is this that makes us fall asleep when we do. If we swamp that rhythm by powerful drugs, that are there not only all night but all day like flurazepam and its metabolites, then the relative contribution of the biological rhythm is swamped, is lost under the influence of the drug. So make sure that people are getting up early if you want them to fall asleep quickly at night.

**Pletscher:** I am interested in this kind of rebound insomnia without a REM rebound, Dr Soldatos. You showed examples of non-hypnotic doses of flunitrazepam, of triazolam, of temazepam nevertheless increasing wake episodes. How do you explain this? Is this compatible with the theory which you mentioned?

**Soldatos:** Yes. The production of endogenous benzodiazepine molecules would be decreased if an active exogenous benzodiazepine drug or compound is introduced. This has nothing to do with effectiveness. The exogenous benzodiazepine drug occupies the receptors and somehow influences either through a feedback mechanism or through the concentration of this exogenous benzodiazepine the production of endogenous benzodiazepine-like molecules. Therefore you have a reduced production of endogenous benzodiazepine-like molecules and it doesn't matter if you have a hypnotic effect or not. After you withdraw the drug, it takes some time until the endogenous production returns to pre-drug levels and this is the critical period when you may have the rebound insomnia.

**Pletscher:** Have you seen this effect with other types of hypnotics?

**Soldatos:** The hypnotic according to our observations has to be a benzodiazepine, a short-acting benzodiazepine not a barbiturate or other drugs. It does not matter how long it is given. It may be given for a short period of time or a long period of time and has not to be withdrawn gradually, but abruptly. This was the observation from the six studies that I mentioned and that was published a couple of weeks ago in *Science*.

**Oswald:** Could you tell us, Dr Soldatos, about that flunitrazepam 1 mg where I think you showed a slide with a 60% increase of wake-time on withdrawal. How many subjects were used in that study?

**Soldatos:** There were twelve subjects, but in other studies we have demonstrated statistically significant changes with only four subjects.

**Oswald:** Some of us would not think that four subjects are susceptible to statistical evaluation, and the reliability of a finding must always come under consideration before we conclude that there are drugs that cause rebound without causing any initial effect.

**Soldatos:** And also the formulation of the drug as Dr Breimer mentioned. In terms of the pharmacokinetics of the drug our subjects are insomniacs, but they are physically healthy subjects; they are usually in the age range of 35 to 50 years, they are checked thoroughly to see that they do not have any physical problems and we do not expect them to have any difference in terms of bioavailability.

**Gaillard:** Several times this morning we have been presented this classification of insomnia, sleep onset difficulties, sleep interruption insomnia and early awakening. In fact, this classification is very old and it has a mainly clinical basis, but in polygraphic studies it has been shown that in fact it is much more difficult to classify the patients according to these criteria. In a recent study comparing insomnia patients we have tried to select patients with a mainly sleep maintenance type of disturbance, and compared to controls we have found that there was no predominance of waking in any part of the night. They had a generally higher trend to waking than the control subjects, but there was no predominance of this excess of waking in any part of the night. Therefore I think this classification of insomnia is questionable.

**Lehmann (Uppsala):** I accept that. I just mean from a clinical point of view you have the impression of different patients complaining about difficulties to get to sleep and they waken too often at night while the endogenous depressed patients wake early. This is my clinical impression and it leads to some implications for using other hypnotic drugs. I give mainly a tricyclic for example for people we would diagnose as having an endogenous depression and no hypnotic drug at all. But I certainly agree with you the classification is much more complicated.

**Gaillard:** You have described a decrease of slow-wave sleep in depressive patients. It has been reported, but we have not found this, and in fact we have found in studying, for instance, insomniac patients with depressed mood a normal amount of slow-wave sleep, comparable to control subjects. It is actually very difficult to score slow-wave sleep visually and this could account for some discrepancy between the studies, but with measurements of the waves you can be more secure in your evaluation of slow-wave sleep.

**Lader:** I am very interested in Dr Lehmann's description of how he came to set up a sleep laboratory. I would just like to mention how we came not to set up a sleep laboratory. We have been interested over the last year or so in monitoring people in real life situations using 24-hour EEG recordings. These are little tiny tape recorders that the person carries round and almost inadvertently we found, of course, with 24-hour recordings we were getting sleep recordings as well and they came through on our computer. I am not in any way trying to belittle the use of sleep laboratories. I think they are extremely important. But I think there are also complementary data which one can get from real life recordings of this sort, and I would like to point out that for every hypnotic tablet that is given in a laboratory there must be hundreds or thousands given in the hospital, and for every one given in a hospital, millions given in the home.

**Lehmann (Uppsala):** This might account for the phenomenon that you can have healthy volunteers in the sleep laboratory and put them on placebo first and then you give them a hypnotic and you find that their time to get to sleep becomes shorter. In the sleep laboratory you might never have real baseline values.

**Pletscher:** Have you any experience with antidepressant drugs? Is REM depression a common denominator for the action of antidepressant drugs? The mechanism of action of antidepressants is more and more questionable. It would be nice to have some common denominator.

**Lehmann (Uppsala):** The only thing I know is that most antidepressant drugs reduce REM sleep. But the remarkable thing is that the tricyclics which are more effective than the MAO inhibitors do so less.

**Pletscher:** Do you have experience with non-tricyclics, non-MAO inhibitors?

**Lehmann (Uppsala):** Yes. I think of tetracyclics. But as far as I know they also have a REM-depressant effect.

**Björqvist:** I was going to raise the same point as Dr Lader. I think a sleep laboratory has its good points. But one thing that is bad about it is that if you have very long registration periods you have the patients sleeping in the laboratory for, let's say, 14 days. That is a very long time. You disrupt several normal patterns of life, for instance, one of the most widely used hypnotic methods, sexual activity, is badly disrupted during such a period. So I think you miss a lot. And then if you have a sleep laboratory and check your patients just during the night, you can never be really sure what they do in the daytime. They might take naps that you will not even know about, for example, sitting in a symposium like this. So you should, if you really want to have good results, have 24-hour records. And I think the only really good method of getting good physiological records is telemetry.

**Soldatos:** I would not dispute the usefulness of 24-hour recordings. We ourselves have already obtained this cassette on which you can record 24-hour activity, but this is a different kind of study. The sleep laboratory has certain advantages as has already been mentioned. You can control the situation whereas you cannot control it at home. Of course, you do deprive the patient of certain activities, it is an artificial environment. But if you compare sleep recorded in the same subjects through telemetry and in the sleep laboratory, it does not differ significantly. At the same time you should think of the subjects being free during the day to pursue their usual activities. In terms of reliability, we are lucky to have subjects in our laboratory that are pretty reliable and whenever they have some extra activity they do tell us, and in that case we can drop this particular night out of the data. But this happens very, very infrequently. We have a special questionnaire where we check all kinds of activities – if they had a common cold, or if they had taken a drug like aspirin or something, or if they had an excess of alcohol. Long-term laboratory studies, of course, provide more difficulties in terms of compliance, as I mentioned in my talk. The shorter the study, the more compliant the subjects are. Of course you make the deal ahead of time, and you make them aware of the possibilities, and if they accept that's fine. Usually you can tell which subject is going to be reliable from your experience, especially if you are a psychiatrist.

**Dworacek (Holland):** Dr Lehmann, you mentioned the use of physostigmine in the case of cholinergic blockade. In anaesthesia we have for many years used physostigmine to counteract the central anticholinergic syndrome after anaesthesia, not only after use of atropine–scopolamine but also in many cases after phenothiazines, and we have seen it in many cases after Rohypnol. We have used it only in the parenteral form. Have you any experience whether it can be used orally?

**Lehmann (Uppsala):** No. I have no personal experience with physostigmine. I just quoted a paper by Sitaram published in 1976 in *Science*. He induced REM sleep by infusing physostigmine in normal volunteers and observed the appearance of REM periods a short time after the infusion. This is in keeping with the view that the cholinergic system might be involved in the timing of REM periods.

**Oswald:** Perhaps I could make a few brief comments. I work in the Poisons Unit at Edinburgh part of the time, where we have 2000 patients a year who have taken overdosages. One of the really big advantages of the benzodiazepines is their safety in overdose. We have an epidemic of overdoses around the world today. This has not yet been mentioned at our symposium. We also had very little mention of the things that help us to sleep naturally. We have had very little discussion about biological rhythms. We have no mention of the role of exercise, for example, if people who are already fit take physical exercise by day they will fall asleep more quickly. They will get more slow-wave sleep in the night and they will have more growth hormone during their sleep. We have heard nothing about the importance of the nutritional state. Those who eat unaccustomed food at bedtime will sleep badly, those who are accustomed to food at bedtime, if they have nothing, will sleep badly. As Crisp in London has shown, people who are losing weight will suffer from broken sleep perhaps, especially in the later part of the night, or will complain of this. People who are gaining weight will sleep more, and this includes people who are depressed. Crisp and his group would claim that it is the change in body weight that is the important thing and not the mood that determines the sleep disturbance. We have heard nothing I think in this meeting about hormones, and there is evidence that in menopausal women, oestrogen replacement therapy will relieve insomnia. We have heard nothing about hormones in relation to benzodiazepines, and one of the really important advances in our knowledge about sleep in recent years has focused on the hormones that are released by sleep, notably growth hormone. This comes specifically with slow-wave sleep, and we have heard from Dr Soldatos about the suppression of slow-wave sleep by benzodiazepines. Well, it is a fact that these substances do not suppress growth hormone. We could, I think, hear more about policy in the use of drugs for the treatment of sleep disorders, the use of antidepressants in particular. We could hear more about the role of age, since Dr Korttila reminded us yesterday about the older brain being more sensitive to benzodiazepines, as others like Castleton and colleagues have shown. We could hear more about the general behavioural effects. The benzodiazepines with Librium are introduced after injection as drugs that would relieve aggressive impulses, but there is good clinical evidence and very good experimental evidence that benzodiazepines release aggression in social situations of frustration. So there are many more topics that we can discuss as well as those our speakers here have drawn our attention to.

**Nicholson:** Dr Soldatos, in the analysis and design of your experiments, am I correct that you compare the effect of a drug, say from days 5–7, with the effect of a placebo given on the previous 2–4 days, and that you do not include in the protocol a placebo study during days 5–7?

**Soldatos:** Let me explain exactly what we are doing. We have one adaptation night, the first night, which is not analysed or statistically evaluated. Then we have three placebo nights, which are the ones with which the other nights are compared. And then we take three drug nights, that is in the 10-night protocol, or seven drug nights in the 14-night protocol. So we evaluate three nights at a time with the three baseline nights, leaving out the first night.

**Nicholson:** This means then that everyone knows in your studies when the drug is being given, since you never give placebo when a drug is being given at the same time? I think this may be important because you see in the withdrawal effect which we have been talking about with triazolam and temazepam, which is a very interesting finding in the first place, the subjects could be aware when the drug is being withdrawn as everyone knows your protocol and you do not have a placebo simultaneously. All your experimenters know, so you are not double-blind, you are single-blind. Another point is that it is only a hypnotic with a persistent effect for several days like flurazepam which would stop these withdrawal effects, whereas short-acting drugs like triazolam and temazepam, which do not last for 24 hours, would not have effects on the withdrawal period. So your subjects could be aware that they are going into withdrawal because you do not use the placebo as a control.

**Oswald:** I think Dr Nicholson has an important point. There are many studies which have shown that placebos will not produce the drug effects on sleep. But there have been none to determine whether withdrawal effects might not arise because the individual is frightened that he may sleep badly once the pill has been withdrawn.

**Lehmann (Zurich):** How many people do you actually use in the individual studies? What is the range? And I also wonder about the selection of the subjects. You said that you prefer to work with insomniacs. Now these people are certainly not normal in the classical sense of the word and there is a wide variety of different causes for their insomnia. In addition, we all know that people who volunteer for more than one night in the laboratory, and even maybe those who volunteer for one night in the laboratory, are very funny people indeed. If you find someone who is willing to go to the laboratory for five nights in a row, this is a most unusual person and I wonder how you shield yourself against a grossly biased sample of the population in this respect?

**Soldatos:** Some studies, as I said before, involve only four subjects. There are other studies where we use up to 12 subjects or even more, depending on the question under study, funds available, and there are many other considerations, including the number of nights. If you have a very long study then obviously the number of subjects is going to be very small. These people are selected as being insomniacs. We prefer insomniacs because they can show the effect clearly. They have a certain amount of sleep difficulties to begin with, therefore, however small, the change is going to show up. If you have normal subjects, you run into the problem that the change is so small that it is undetected finally. Testing hypnotics in non-insomniacs is like testing antidepressants in non-depressed individuals. Insomniacs are the people that take hypnotics, therefore they are the best subjects for this type of study. Now, of course, they are not normal and, of course, insomnia has many causes, but we try to select subjects who are physically fit and also do not have major emotional or psychiatric problems. Of course, it is not a homogenous population because of the many causes of insomnia, but most of them are monosymptomatic.

213

**Björqvist:** We are very technical now, I think, but this shows we have noticed some points concerning Dr Soldatos' protocol. First of all, there are too few subjects, there are no real placebo nights and there is no exact daytime control of napping for instance. And finally, you get a peculiar selection of volunteers. I wonder how one can get out of these difficulties without making a sleep laboratory too big. Your volunteers are mostly the same people coming again and again for testing different drugs. Is it really necessary to get a baseline night every time and also to avoid first-night effects?

**Soldatos:** We use new people in each study, so it is not a problem of using the same people. As to the baseline nights not being necessary, I do not agree. Just having one night every week, or something like that, has many, many disadvantages: you do not know what the continued effect of the drug would be. In the real situation, these drugs are taken for many nights and not just for one night, especially with chronic insomnia, and also you run into the problem of having many different influences; you may have all kinds of other drugs taken in between that may affect your drug nights.

**Erdmann:** As an anaesthesiologist I am concerned with putting people to sleep and giving them an analgesic at the same time. One thing I have missed, and which was not mentioned in this symposium, is the question of respiratory side-effects induced by hypnotics like diazepam. You just cannot combine them or increase the dosage as you want. You will have respiratory distress and the respiratory rate will decrease, which means the $pO_2$ values would fall and $pCO_2$ values would go up with all the problems of intercranial pressure increase. In anaesthesiology we use ketamine, which is a very good analgesic, but it has side-effects in the form of bad dreams. If we combine it with diazepam as sedative drug we can reduce the dosage and antagonize or counteract the side-effects of ketamine. If you have 250 mg of ketamine and 50 mg of Valium mixed in 500 cc solution, and if you need to give 250 mg of ketamine in 2 hours, it means the patient also gets 50 mg of Valium which is quite a high dosage. Alternatively you can give it with a single dose of diazepam, which is much more appropriate because diazepam has a long action whereas the analgesic action of ketamine is shorter.

**Monti:** I think it has to be emphasized that the withdrawal insomnia mentioned by Dr Soldatos is very frequently related to REM rebound. Using polygraphic recordings it can be shown that there are two kinds of REM rebound. The first is obtained even with relatively small doses, as we found with 1 mg of flunitrazepam in normal volunteers, and is associated with an increase of REM sleep during the later part of the night. But the other one, which is much more important, appears after the administration of larger doses of flunitrazepam and associated with the presence of nightmares, which I believe could be responsible for the continuous use of hypnotics in this kind of patient.

214

**Soldatos:** There is certainly a relation between REM withdrawal, REM rebound and rebound insomnia. According to our experience, this is true for the drug withdrawal insomnia described by Dr Kales many years ago with the barbiturates, and in that case the dosage of the drug was multiple, people and took the drug for many months or years, sometimes in multiple doses every night. When the drug was withdrawn abruptly, a REM rebound occurred, and also insomnia, which was named drug withdrawal insomnia. But we have not seen this in our current studies with benzodiazepines.

**Gottfries:** We know that some systems in the brain are weak in the elderly: for instance, anticholinergic drugs can have a rather serious effect in older people. We have a central anticholinergic syndrome in old age. We heard yesterday that the effects of the benzodiazepines increase in the elderly. In some investigations of mine I found that biologically old patients get a very serious atonic ataxia and sometimes confusion. This syndrome appears sometimes a week after starting treatment and it may be because it takes so long to reach the steady state in these patients. I would very much like to provoke the auditorium here by asking how we should treat aged people. Can we recommend benzodiazepines to aged people? I in fact am very careful about this and recommend them only with complete follow-up and careful supervision of these patients, to ensure these toxic syndromes will not appear. I should like to ask here if it is possible to study this experimentally, as we know that the age changes in the rat, for instance, resemble those in the human brain. Are there changes in the benzodiazepine receptors in older animals? Can that question be answered today?

**Pletscher:** I cannot answer all Professor Gottfries' questions, but I can say that benzodiazepines with regard to sleep are somehow different in animals than in man. On the other hand, it is not yet sure that the so-called benzodiazepine receptor is connected with sleep. It could also be related to muscle relaxation or to something else, but there are many problems. Another question is whether there is an influence of food on the quality of sleep. You know that proteins contain amino acid precursors of biogenic amines, tryptophan, phenylalanine, and it is also known that by loading tryptophan you can increase serotonin in the brain, which may be related to sleep. Now what should one eat? Should one eat a juicy American steak before going to bed, should one eat English porridge or Swiss Rösti, or what would you recommend?

**Oswald:** Milk and cereal. But more important, that which you are accustomed to take.

**Pletscher:** You would not bother about these amino acid precursors? Of course, amino acids can have the opposite effect in that they competitively inhibit the precursors from entering the brain. But is there no knowledge as to whether high protein meals would be better for insomniacs?

215

**Oswald:** It is not yet established anyway. It's better to take what you usually take; any departure from the normal will disturb your sleep.

**Lehmann (Uppsala):** In a book by Professor Wyatt about sleep disorders he says a good steak contains between 0.5 g and 2 g of tryptophan and that should act as a good sedative.

**Oswald:** But it contains a lot of other things as well, including a lot of fat.

**Bartholdson (Sweden):** Have any of the speakers experience of hypnosis or meditation in the sleep laboratory? *(No response.)*

**Lingjærde:** We performed a controlled clinical trial on the effect of nitrazepam in relation to age in Norway in 1978 *(Curr. Ther. Res.,* **24,** 388) (Figure 1). This was a trial on 61 general practice patients with insomnia who were randomized on nitrazepam 5 mg or alimemazine (trimeprazine)

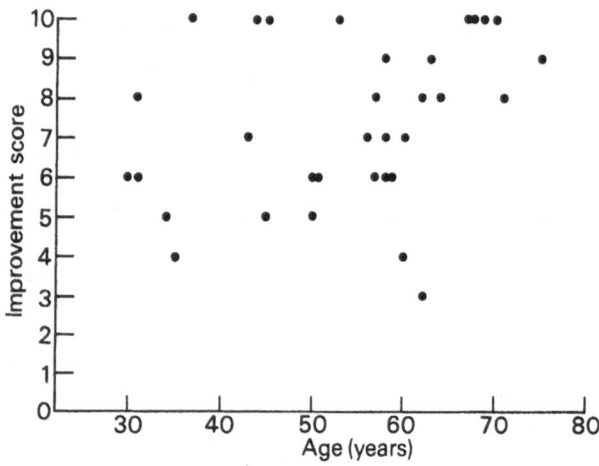

**Figure 1** Global improvement of sleep on nitrazepam *vs* age. $r = 0.31, 0.10 > p > 0.05$

20 mg at night. The treatment period was 1 week, with another 2 or 3 weeks for those who wanted to continue. Overall in this trial nitrazepam turned out to be a more effective hypnotic, especially during the first week, and it seemed to have a specially favourable effect in the patients above the age of 65. By chance patients on this age numbered 60 in nitrazepam group but only one in the alimemazine group, so these two hypnotics could not be compared in elderly patients. Figure 1 shows global improvement of sleep in the first treatment week in the nitrazepam group of 73 patients in relation to age. The global improvement score is derived by comparing the score for the total assessment of sleep in the first week to sleep before the

trial. The highest possible score on this measure is ten and the lowest is two. A linear correlation analysis of the data shown in the figure gives a correlation coefficient of 0.31, which is not quite significant $(0.10 > p > 0.05)$. However, from these data one can safely draw the conclusion that nitrazepam is at least as effective in elderly patients as in younger aged groups and possibly better. It should be added that the frequency of side-effects was low in the total patient group and did not seem to be higher in older than in younger patients. This seems to be in accordance with the large material which was recently published from the Boston Collaborative Drug Surveillance Program, demonstrating that, in a dosage of 5 mg, nitrazepam has a rather low toxicity in elderly patients.

**Gottfries:** I also think that these drugs as anti-anxiety drugs or hypnotic drugs have the same good effect in old people as in young people, and this is not the reason that I raised the question whether we should use them or not. It is just that side-effects do appear after a couple of days' use or 1 week and I therefore would like to ask Professor Lingjærde if these patients were on continuous treatment for a longer time until you reached the steady-state level, and if there were then serious side-effects in this group, because that is what I think is dangerous and that may complicate the treatment in older patients.

**Lingjærde:** We did not follow the patients for more than 3 or 4 weeks. There was no increasing frequency of side-effects during this period, but I have some experiences with Rohypnol for up to 2 years in older people, especially two older men I know, both of them about 80 years old, who have used 1 mg of Rohypnol almost every evening for at least 2 years and the effect seems to be very good throughout; they have no complaints or any type of side-effects so I think in this dosage it is very safe and very effective, even in such old patients.

**Wickstrøm:** I also used flunitrazepam 1 or 2 mg in about 12 patients between 70 and 78 years for 6 to 12 months and the effect was exactly the same as with the younger ones, and also the side-effects.

**Soldatos:** According to a paper published recently in the *Journal of the American Medical Association*, 75% of the geriatric patients that are taking psychotropic drugs are using hypnotics. This is a very large percentage, and if you think that 25% of all people are taking psychotropic drugs, you have a very large number of people. We have always to keep in mind the side-effects of these drugs and also such things as rebound insomnia.

**Oswald:** I should just like to remind you that although flurazepam may continue to give you increased sleep as Dr Soldatos showed when it is being withdrawn, this is really because it is persisting in the body and when you take it regularly at night for 3 weeks, in your third week if you measure skilled performance, whether it is vigilance or card sorting or manual dexterity or here simple digit substitution, there is impairment as a con-

217

sequence of taking flurazepam 30 mg at night. There is impairment in the morning, in the afternoon and in the evening. We studied 12 subjects in a cross-over design having 2 hours of testing in the morning, 2 hours of testing in the afternoon and 2 hours of testing in the evening. This is just to remind you that some of the effects of flurazepam that may be good in some ways can be bad in others.

**Gaillard:** I would like to raise the problem of the effect of benzodiazepines on slow-wave sleep. There are two aspects in these effects. First of all, it has been repeatedly shown that benzodiazepines markedly depress slow-wave sleep, but I am not aware of a single study which definitively establishes whether or not there is a rebound of slow-wave sleep. In other circumstances, e.g. after instrumental deprivation of slow-wave sleep, it has been shown that there is a rebound of slow-wave sleep. Secondly, there is the fact that benzodiazepines depress slow-wave sleep, whereas they do not depress the peak of growth hormone secretion.

**Soldatos:** I did not find in any of our studies that benzodiazepines produce a rebound of slow-wave sleep. The studies I mentioned are not of benzodiazepines but of barbiturates. As Dr Oswald mentioned, there have been studies that show that there is a dissociation between the sleep-stage effects and the effects on the hormones.

**Gaillard:** I think this is important. Have we to admit that slow-wave sleep is dissociated and only some components of it are affected by benzodiazepines, or is there some other explanation?

**Soldatos:** I do not know. I have in mind the study by Feinberger published recently in the *EEG Journal* which showed that the slow waves in stage II are more numerous and finally offset the loss of delta waves in stages III and IV.

**Lehmann (Uppsala):** We have 3 subjects who showed an increase of slow-wave sleep about 3 days after 25 mg of oxazepam. But we did not believe these results and we have not published them yet.

**Hartelius:** We had a group of 119 subjects in whom we studied clinically the withdrawal of flunitrazepam and also phenothiazines and so on. And we expected that we should have many complaints from the patients because they had used it for several months. It was quite possible to perform this withdrawal and they did not complain very much, and the majority agreed to a repeat examination. Nevertheless, most of them returned to their old drug, either benzodiazepines or phenothiazines, and we draw the conclusion that with the prolonged use of hypnotics the pharmacological effect decreases and is replaced by a sedative placebo effect. I should like to ask Dr Soldatos: did your patients complain about rebound insomnia? It is rather important to have this clinical assessment.

**Soldatos:** In our study, because our measurements are so precise and accurate, we can detect rebound insomnia, and this is probably the reason why your study based on clinical criteria did not show it. In terms of their subjective feelings, yes, there was a worsening of sleep, but not as much as our objective findings.

**Monnier:** Dr Oswald, I very much appreciated your remarks about age and bodyweight. I think there is a gap between the importance that we as experimenters, physiologists or pharmacologists give to the factor bodyweight and the tendency to minimize this factor in therapeutic dosage of drugs and clinical pharmacology. I know, for instance, what occurred when Rohypnol was introduced. It was introduced in Europe, in our country, with a strong recommendation for a low dosage of 1 mg. Some 6 months later the journals from the United States were saying 1 mg is not enough for American patients, we need 2 mg. I think what you said about sensitiveness to benzodiazepines in age could be explained by changing bodyweight. I think in the clinical presentation of the drugs this is not considered any more. Dosage is usually standardized for a bodyweight of 70 kg, and then we have always the impression that the dose which is recommended in the States as optimum does not fit in Europe because of differences in the bodyweight. If we take the case of the older person, he loses bodyweight, bodylength (3 cm at least in the vertebral column), his brain loses water. Now you give the same dosage to these old people without considering and correcting for bodyweight. And what are the important targets? The important targets are the neuron and the retinal cell. Professor Pletscher may correct me, but I do not think that the influence of benzodiazepines has been studied on the retina. According to my calculations I think that we very often give older people doses between one-quarter and one-third too high. Anaesthesiologists do consider the bodyweight, of course, and paediatricians too, but it should also be considered for the use of benzodiazepines in general practice.

**Oswald:** Professor Monnier and I both had the privilege to be at the Roche symposium in Zurich in 1964 and I think we can both see what a long way the benzodiazepines have come in that time. Professor Hess was present too and we have passed through an era when people thought about the reticular formation and little else when they considered sleep. We have gone through the era of the biogenic amines and little else, and that's passing too, and I think we are going to be moving on to the era when we consider the restorative function of sleep and the effects of drugs on that function. This was one of the things that Hess emphasized in that first meeting and that has still not yet been fully considered.

**Pletscher:** Ladies and gentlemen. We have come to the end of our one-and-a-half days' symposium on sleep research. From our standpoint it has certainly been a successful meeting. You have covered a lot of ground regarding theoretical and clinical aspects, although many topics have been

219

left out, as today's chairman has mentioned. We have been given insight into the fascinating field of neurotransmitters and the central nervous system and receptors which may be involved in the regulation of sleep and in the action of drugs on sleep. The fields of neurotransmitters and their receptors have undergone an explosive development in recent years, and this has also reflected on sleep research. On the other hand, the accumulation of knowledge is such that its integration into a comprehensive concept appears to be a monumental task. One has almost the impression that the brain is too complex a structure to be fully explored by the human brain. We have also been shown the fascinating field of regional blood flow and metabolism of the brain *in vivo*. The development of new methodology makes this very difficult field accessible for research and I have the feeling that many new findings of both practical and theoretical interest will come out of this type of research. In this connection it was also rewarding to see that whereas higher intellectual activity up to now has been thought not to need relevant amounts of metabolic energy, clear-cut changes have been shown with this technique. Whether relevant amounts of metabolic energy can be diverted to the brain, and whether hard thinking may develop into a cure for obesity, I cannot predict.

The sessions on clinical pharmacology, including treatment of sleep disorders, were mainly concerned with benzodiazepines. Progress has been made here. Problems of pharmacokinetics, of the pharmacological spectrum, and of the quality of sleep were in the forefront of our interest. It was most useful to hear that in order to improve mental performance next day, you should go and see horror films in the evening. I hope that this finding will not become known to the TV people! We have also heard from today's chairman that the old English saying 'Early to bed and early to rise makes a man healthy, happy and wise' has finally received a scientific basis. It was also rewarding to see that the methodology of testing drugs in sleep disorders has greatly improved, e.g. with the introduction of sleep-laboratory techniques. There was some controversy in this discussion, but I can tell you that we have greatly benefited from this type of research which was able to differentiate between the actions of the various derivatives of the benzodiazepine group. Despite all this progress, the problems to be elucidated are far more numerous than those solved. The exact role of neurotransmitters in regulating sleep, the role of GABA-ergic and other systems in the sleep-inducing actions of benzodiazepines, the differences in the pharmacological spectrum and in the quality of action between the benzodiazepines and the older hypnotics, the mechanism of action of benzodiazepines in man and many others require further clarification. Sleep remains a fascinating topic for basic and clinical research. I hope that this symposium may act as a new catalyst for your personal experimental and clinical work. I wish you further success, a little bit of luck, which is always needed, and happy return to your homes. Thank you very much.

# Index

acebutol, 25
acetylcholine, 10, 19, 29–30, 201
$a$- and $\beta$-adrenergic receptors, 19–30, 38
agonist–antagonist interactions, 26–7
allypropymol *see* aprobarbital
amnesic action of benzodiazepines, 123–7
antidepressants, effect on sleep, 40, 50, 52–3, 210
aprobarbital, compared with flunitrazepam, 157–8, 166–8, 205
atropine, 29

barbiturates
  effects of, 149–50, 168, 176
  elimination half-lives of, 66
benzodiazepine
  amnesic action and residual effects of, 123–31
  clinical effects and concentrations, 99–107, 146–8
  effect on slow wave sleep, 52, 218
  effects on sleep and performance, 109–21
  mechanism of action, 3–11, 55
  pharmacokinetics of, 73–9
  receptor, in brain, 3–11, 45, 55, 105, 190
  sleep laboratory studies on, 184–94
  treat sleep disorders, 172–5, 177, 204–5
  *see e.g.* diazepam, flunitrazepam *etc.*
binding site for benzodiazepines, 3–7
blood flow, cerebral, 13–17, 46, 47–9
brain
  benzodiazepine receptor in, 3–11, 45, 55, 105, 190

blood flow and metabolism during sleep, 13–17, 46, 47
catecholaminergic systems, 35–40
drug penetration into, 67
butobarbital, 66, 70

catecholamines, cerebral, 20, 30, 35–40
CBF, 13–17
cerebral circulation and metabolism in sleep, 13–17, 46–9
chloral hydrate, 173, 176–7
chlorpromazine, 29, 38–9, 177
chlorpropamide, hypersomnia with, 199
clobazam, effect on sleep and performance, 109, 114–15
clonidine
  effect on PS, 37–8
  effect on receptors, 46
  effect on sleep stages, 21–3, 26–30
clorazepate, effect on sleep and performance, 109, 111–14
cortex, cerebral, benzodiazepine affinity in, 45
cremophor EL, solvent for diazepam, 130

diazepam
  amnesic action of, 125–6
  binding of, 3–4
  effect on sleep and performance, 116–21
  metabolites of, 73–4, 103–4
  pharmacokinetics of, 76–7
  relation between effect and concentration, 102, 106, 148
  residual effects of, 127–30
dibenamine, 27

dreams
  after stress, 139–40
  relation to blood flow landscapes, 54
drugs, hypnotic *see* hypnotic drugs

elimination half-life of drug, 64, 66,
  73–4
  *see also individual drugs*
emission tomography, 13

flunitrazepam (Rohypnol)
  amnesic action of, 126
  clinical effects of, 88–96, 148, 186
  compared with other hypnotics,
    156–69, 205–7
  elimination half-life of, 97–8
  pharmacokinetics of, 83–8
  residual effects of, 127–30
flurazepam
  compared with flunitrazepam, 164–6
  effects of, 188–9
  pharmacokinetics of, 76
fosazepam, effect on sleep and
    performance, 109, 111–14

GABA receptors, 9–10, 45, 55
global cerebral blood flow (CBF),
  13–17

half-life, elimination, 64, 66, 73–4
heptabarbital, 66, 70
hexobarbital, 66
Huntington's disease (chorea),
    alteration of benzodiazepine
    receptors in, 3, 9, 46, 58
3-hydroxydiazepam *see* temazepam
6-hydroxydopamine (6-OHDA)
  effect on PS, 36, 40
  effect on wakefulness, 35
5-hydroxytryptamine
  effect of benzodiazepines on, 10
  effect of α- and β-receptors on, 30
  role in sleep, 178
hypnotic drugs
  comparative studies, 155–69
  duration of action, 63–71
  misuse of, 203–4
  onset of action, 70–3
  testing in sleep laboratory, 181–94
  use of, 171–8

*see also* benzodiazepines, *individual
  drugs*

'intrasleep rebound', 38

ketamine, 150, 214

LC *see* locus caeruleus
locus caeruleus, in PS, 19, 28–9, 38
lorazepam
  amnesic action of, 126–7
  residual effects of, 129

Mandrax *see* methaqualone-diphen-
    hydramine
melanocyte-stimulating hormone, 46
melatonin, serum, relation to sleep,
    43–4, 47, 53
methaqualone-diphenhydramine, 176
  compared with flunitrazepam, 162–4
methoxamine, 30
α-methyldopa, effect on sleep stages,
    21–2, 27
α-methylparatyrosine (αMPT)
  effect on PS, 39
  effect on wakefulness, 35
  inhibits CA synthesis, 37
  inhibits NE synthesis, 19–20
metoprolol, 25
mianserin, 50
midazolam maleate, 131
Mogadon, 75
mood, relation to sleep quality, 135–6,
    141–3
αMPT *see* α-methylparatyrosine

NE *see* norepinephrine
neurons
  GABAergic, benzodiazepine
    receptor on, 9
  raphe, 30, 178
neurotransmitter, 9–10
  *see e.g.* acetylcholine, serotonin
nicotinamide, prolongs total sleep and
    REM sleep, 10
nitrazepam,
  compared with flunitrazepam,
    158–61
  effect in relation to age, 216
  half-life, 70

pharmacokinetics of, 74 6
'non-barbiturates', 176
nordiazepam, effect on sleep and
    performance, 109, 111–14
norepinephrine (NE)
    role in mechanism of sleep stages,
        19–30
    role in waking, 35

6-OHDA *see* 6-hydroxydopamine
oxazepam
    effect on sleep and performance,
        109, 116–21
    pharmacokinetics of, 78

paradoxical sleep (PS)
    EEG in, 21
    effect of $\alpha$- and $\beta$-receptors on, 20–30
    role of brain CA systems, 35–40
    role of noradrenaline in, 19
Parkinson's disease, 36
pentobarbital, 188–9
phenoxybenzamine, effect on sleep
    stages, 23–4, 27, 28, 30
phentolamine, effect on sleep stages,
    23–4, 26–9
physostigmine, shortens REM latency,
    201, 211
pindolol, effect on sleep stages, 25
piperoxane, 24, 27, 30
propranolol, effect on sleep stages,
    25–7, 30
PS *see* paradoxical sleep

questionnaire, for sleep quality, 143–4,
    183

raphe neurons, 178
    effect of clonidine on, 30
rCBF, 13–17
'rebound insomnia', 189–90
receptor
    benzodiazepine, in brain, 3–11, 45,
        55, 105, 190
    $\alpha$- and $\beta$-, 20–30, 38
regional cerebral blood flow (rCBF),
    13–17
REM sleep
    cerebral blood flow in, 48–9

effects of benzodiazepines on, 112,
    118
effects of drugs on, 176, 191
effect of $\alpha$- and $\beta$-receptors on, 23–5,
    29
increased by nicotinamide, 10
melatonin levels in, 44
suppression of, 52–3
use in diagnosis, 199–201
reserpine, suppresses PS, 35–6, 40
residual effects of drugs, 123–5, 127–30
Rohypnol *see* flunitrazepam

saliva, concentration of drug in, 105–6
seasonal variations in sleep, 43–4, 57–8
secobarbital, 66, 188
sedation with intravenous benzo-
    diazepines, 123–31
serotonin
    in PS, 19
    role in sleep, 178
sleep
    brain CA systems in, 35–40
    cerebral circulation and metabolism
        in, 13–17
    disorders, 171–5
    effect of benzodiazepines on, 109–21
    effect of nicotinamide, 10
    effect of $\alpha$- and $\beta$-receptors on, 19–30
    factors disturbing, 137–41, 149, 151,
        200
    -inducing drugs *see* hypnotics,
        benzodiazepines *etc.*
    laboratory, role of, 181–94, 197–202,
        210
    paradoxical *see* paradoxical sleep
    quality, measurement of, 135–44,
        149; questionnaire, 143–4
    REM *see* REM sleep
    serum melatonin during, 43–4, 47,
        53
    slow-wave *see* slow-wave sleep
    stages, 20, 190–3
slow-wave sleep
    brain metabolism in, 14–17
    EEG in, 21
    effect of drugs on, 191–2, 218
    effect of $\alpha$- and $\beta$-receptors on, 20–30
    use in diagnosis, 200–1
Steele–Richardson disease, 36

temazepam
  effect on sleep and performance,
    109, 116–21, 188–9
  pharmacokinetics of, 77, 78, 149
thymoxamine, effect on sleep stages,
    23–4, 28
tolazoline, effect on sleep stages, 23–4,
    26–7, 28
triazolam
  effects of, 188
  pharmacokinetics of, 78–9
  triflubazam, effect on sleep and
    performance, 109, 114–15
L-tryptophan, effect on sleep, 178

wakefulness
  brain CA systems in, 35
  brain metabolism in, 16
  effect of α- and β-receptors on, 20–30
  role of noradrenaline in, 19

vigilance, states of, 20–1

xylazine, effect on sleep stages, 21–3,
    27, 28

yohimbine, effect on sleep stages,
    23–4, 26–30